Symmetry and

Eric Lord is a British mathematician with a PhD from King's College, London (1967), resident in India. From 1984 to 2006 he was associated with the Indian Institute of Science, Bangalore. His research interests have been mainly in general relativity and algebraic structures in elementary particle physics, and latterly in geometrical structures in materials science: crystallography, quasicrystals, minimal surfaces.

Other Books by Eric Lord

Tensors, Relativity & Cosmology. Tata McGraw-Hill, Delhi, 1976
E A Lord & C B Wilson. The Mathematical Description of Shape and Form. Ellis Horwood, 1984
E A Lord, A L Mackay & S Ranganathan. New Geometries for New Materials. Cambridge University Press, 2006
Science, Mind and Paranormal Experience. Lulu.com, 2006

Eric Lord

Symmetry and Pattern in Projective Geometry

Eric Lord
Bangalore, India

ISBN 978-1-4471-4630-8   ISBN 978-1-4471-4631-5 (eBook)
DOI 10.1007/978-1-4471-4631-5
Springer London Heidelberg New York Dordrecht

Library of Congress Control Number: 2012954673

Mathematics Subject Classification: 00A66, 14N10, 14N20, 51A45, 51A05, 51E15, 51E20, 51N15, 97K26

1st edition © Eric Lord 2010, published by Springer-Verlag London 2013. All Rights Reserved.

The Annunciation by Fra Carnevale (p. 3). This work is in the public domain in the United States, and those countries with a copyright term of life of the artist plus 100 years or less.

Man Drawing a Lute by Albrecht Dürer (p. 4). This work is in the public domain in the United States, and those countries with a copyright term of life of the artist plus 100 years or less.

© Springer-Verlag London 2013
This work is subject to copyright. All rights are reserved by the Publisher, whether the whole or part of the material is concerned, specifically the rights of translation, reprinting, reuse of illustrations, recitation, broadcasting, reproduction on microfilms or in any other physical way, and transmission or information storage and retrieval, electronic adaptation, computer software, or by similar or dissimilar methodology now known or hereafter developed. Exempted from this legal reservation are brief excerpts in connection with reviews or scholarly analysis or material supplied specifically for the purpose of being entered and executed on a computer system, for exclusive use by the purchaser of the work. Duplication of this publication or parts thereof is permitted only under the provisions of the Copyright Law of the Publisher's location, in its current version, and permission for use must always be obtained from Springer. Permissions for use may be obtained through RightsLink at the Copyright Clearance Center. Violations are liable to prosecution under the respective Copyright Law.
The use of general descriptive names, registered names, trademarks, service marks, etc. in this publication does not imply, even in the absence of a specific statement, that such names are exempt from the relevant protective laws and regulations and therefore free for general use.
While the advice and information in this book are believed to be true and accurate at the date of publication, neither the authors nor the editors nor the publisher can accept any legal responsibility for any errors or omissions that may be made. The publisher makes no warranty, express or implied, with respect to the material contained herein.

Springer is part of Springer Science+Business Media (www.springer.com)

There are many geometries, each describing another world: wonderlands and Utopias, refreshingly different from the world we live in.

H S M Coxeter

Preface

The methods and principles of Projective Geometry have a unique elegance. Intricate and surprising structures unfold from a very few, very simple concepts. It has been among my favorite mathematical interests for a very long time. I wrote this book because I wanted to convey to others some of the fascination I feel for this subject. It is not a 'textbook'. It is a collection of ideas that have especially appealed to me, presented in a way that emphasizes general principles and that, I hope, will make you want to explore further. I have tried to avoid abstruse terminology and esoteric notation as much as possible, so it should be accessible to anyone with a little knowledge of matrices, determinants and vectors, and perhaps a smattering of group theory.

The characteristic of geometry that distinguishes it from other kinds of mathematics such as algebra or number theory, and that in a sense serves as a definition of 'geometry', is that it appeals to the imagination through intuitions arising from the way we perceive the 'real world'. This remains so even when geometrical ideas wander into realms of higher dimensions and structures that are hard—or impossible—to 'visualize'. So long as the link to spatial intuitions remains, however tenuous, we are still in the land of Geometry. Once that link is gone, it is no longer Geometry, it is Algebra—where the patterns and structures are abstract and have a different kind of appeal. But there is no sharp boundary.

Projective Geometry grew out of the efforts of architects and painters to represent the three-dimensional world on a flat two-dimensional surface. This is not a question of the geometry of the shapes of things as they 'really are', but rather the geometry of how they 'seem to be' when we look at them. Through the efforts of many eminent mathematicians this geometry of 'perspective' was developed into a whole new and exciting branch of mathematics; my aim in writing this book is to convey something of the flavor of this development, which reached its highest point in the 19th century and the first half of the 20th century. Thereafter, interest declined and projective geometry (and indeed, geometry in general) came to be regarded as quaint and 'old-fashioned', as mathematicians became more and more interested in abstractions. Happily, the tide seems to be turning, probably as a result of the advent of efficient and versatile computer graphics systems which are re-establishing the link between mathematics and visual perception.

The greatest geometer of the 20th century was H S M 'Donald' Coxeter. Throughout an exceptionally long life he produced a torrent of beautiful and novel geometrical ideas, all expressed with inimitable elegance. All mathematicians—including me—who maintained an active interest in geometry throughout its 'unfashionable' period are indebted to him.

That's all. Hope you enjoy my book.

Bangalore, India   Eric Lord

Contents

1 Foundations: The Synthetic Approach 1
   1.1   Euclid 1
   1.2   Axioms of Projective Geometry 2
   1.3   The Art of Perspective 3
   1.4   Desargues' Theorem 5
   1.5   The Complete Quadrilateral 7
   1.6   Affine Geometry 10
   1.7   The Theorem of Pappus 11
   1.8   Affine Coordinates 13
   1.9   Configurations 18
   1.10 Axioms for N-Dimensional Projective Space 24
   1.11 Duality 25
   1.12 Algebra *versus* Axioms 25

2 The Analytic Approach 27
   2.1   Homogeneous Coordinates 27
   2.2   More than Two Dimensions 30
   2.3   Collineations 31
   2.4   A Proof of Pappus's Theorem 33
   2.5   Proofs of Desargues' Theorem 34
   2.6   Affine Coordinates 36
   2.7   Subspaces of a Vector Space 36
   2.8   Plücker Coordinates 38
   2.9   Grassmann Coordinates 39

3 Linear Figures 43
   3.1   The Projective Line 43
   3.2   Cross-Ratio 45
   3.3   Involutions 46
   3.4   Cross-Ratio in Affine Geometry 47
   3.5   The Complex Projective Line 48

	3.6	Equianharmonic Points	50
	3.7	Four Points in a Plane	50
	3.8	Configurations in More than Two Dimensions	53
	3.9	Five Points in 3-Space	54
	3.10	Six Planes in 3-Space	55
	3.11	Six Points in 4-Space	58
	3.12	Sylvester's Duads and Synthemes	60
	3.13	Permutations of Six Things	62
	3.14	Another Extension of Desargues' Theorem	63
	3.15	Twenty-Seven Lines	64
	3.16	Associated Trihedron Pairs	66
	3.17	Segre's Notation	68
	3.18	The Polytope 2_{21}	68
	3.19	Desmic Systems	69
	3.20	Baker's Configuration	72
4	**Quadratic Figures**	79
	4.1	Conics	79
	4.2	Tangents	81
	4.3	Canonical Forms	82
	4.4	Polarity	84
	4.5	Self-polar Triangles	86
	4.6	Mutually Polar Triangles	88
	4.7	Metric Planes	89
	4.8	Pascal's Theorem	92
	4.9	The Extended Pascal Figure	95
	4.10	Quadrics	99
	4.11	Pascal's Theorem Again	100
	4.12	Tangent planes	101
	4.13	Polarity	102
	4.14	Affine Classification of Quadrics	103
	4.15	Reguli	104
	4.16	Metric Spaces	107
	4.17	Clifford Parallels	108
	4.18	Isometries	110
5	**Cubic Figures**	115
	5.1	Plane Cubic Curves	115
	5.2	Nine Points	117
	5.3	A Canonical Form for a Plane Cubic Curve	117
	5.4	Parametric Form	118
	5.5	Inflections	119
	5.6	Cubics in Affine Geometry	120
	5.7	The Twisted Cubic	121
	5.8	Chords, Tangent Lines and Osculating Planes	122
	5.9	A Net of Quadrics	124

	5.10 Cubic Surfaces	125
	5.11 Canonical Forms for a Cubic Surface	128
	5.12 Twenty-Seven Lines	129
6	**Quartic Figures**	**133**
	6.1 Algebraic Geometry	133
	6.2 The Hessian of a Cubic Surface	134
	6.3 Desmic Surfaces	136
	6.4 Kummer's Quartic Surface	138
7	**Finite Geometries**	**145**
	7.1 Finite Geometries	145
	7.2 PG(2, 2)	146
	7.3 PG(2, 3)	147
	7.4 PG(3, 2)	148
	7.5 Galois Fields	148
	7.6 PG(2, 4)	150
	7.7 Structure of PG(N, q)	151
	7.8 Collineations of PG(N, q)	154
	7.9 Finite Projective Lines	155
	7.10 PL(3)	156
	7.11 PL(5)	156
	7.12 Six points in PG(2, 4)	159
	7.13 PL(7)	160
	7.14 Eight Points in PG(3, 2)	161
	7.15 Steiner Systems	163
	7.16 PL(11)	163
	7.17 Coxeter's Configuration (11_6)	163
	7.18 Twelve Points in PG(5, 3)	165
	7.19 PL(23)	166
	7.20 Twenty-Four Points in PG(11, 2)	167
	7.21 Octastigms and Dodecastigms	169
Appendix	**Notes and References**	**173**
Bibliography		**179**
Index		**181**

Chapter 1
Foundations: The Synthetic Approach

Abstract The axioms for Euclidean, Affine and Projective geometries in two and three dimensions are discussed, with particular emphasis on the status of the theorems of Pappus and Desargues. The origin of Projective geometry in the art of 'perspective drawing' is explained. The arithmetic satisfied by the points of a line in affine space is demonstrated and shown to lead to the possibility of affine coordinate systems. After a digression on the symmetry groups of configurations, a set of axioms for Projective geometry of N dimensions is presented.

1.1 Euclid

The word 'geometry' derives from the Greek $\gamma\varepsilon\omega$ (Earth) $\mu\varepsilon\tau\rho\iota\alpha$ (measure)—indicating the origins of this branch of mathematics in the methods of land surveying. As a *practical* branch of mathematics it is of course the knowledge needed by architects, engineers, carpenters—and land surveyors. The emphasis is then on the science of *measurement*—the study of properties of angles, lengths, areas and volumes. In the third century BC Euclid of Alexandria systemized the subject, logically deriving an abundance of geometrical results from a list of 'definitions' (introducing the concepts of points, lines, planes, and metrical concepts including lengths of line segments, angles, circles, *et cetera*) followed by five fundamental 'postulates' (axioms). The thirteen books of Euclid's *Elements of Geometry* constitute the most remarkable mathematical work ever. It still provides the foundations of the subject after 23 centuries.

The first four postulates are simple and might be regarded as 'self-evident':

I. A straight line segment can be drawn joining any two points.
II. Any straight line segment can be extended indefinitely in a straight line.
III. Given any straight line segment, a circle can be drawn having the segment as radius and one endpoint as center.
IV. All right angles are congruent.

The fifth postulate is less simple and became a source of controversy:

V. If a straight line falling on two straight lines makes the interior angles on the same side less than two right angles, the two straight lines, if produced indefinitely, meet on that side on which the angles are less than two right angles.

Two lines in the same plane that do not intersect, however far extended, are said to be *parallel*. A more useful and more elegant postulate, which can be shown to be equivalent to Euclid's fifth postulate, is 'Playfair's axiom':

V'. Given a plane containing a line and a point not on the line, there is a unique line parallel to the given line through the given point.

Many mathematicians tried in vain to derive the notorious fifth postulate as a theorem—a logical consequence of the other four. It was not until the 19th century that Bolyai and Lobachevsky showed conclusively that the fifth postulate could be thrown out and replaced by the assumption that there could be an infinite number of lines through a given point and 'parallel' to a given line, without losing logical consistency. They had discovered a *non-Euclidean* geometry.

1.2 Axioms of Projective Geometry

An alternative approach is to throw out the concept of parallel lines altogether. If we take the drastic step of also throwing out all concepts of measurement (no angles, lengths, areas—no *metrical* concepts at all), then only *incidence* properties remain: a point and a line are *incident* if the point lies on the line. A point (or a line) and a plane are *incident* if the point (or the line) lies in the plane. (For ease of expression we frequently rephrase such statements in other ways: a line or a plane may be said to *pass through* or *contain* a given point, and so on.)

Thus we are dealing essentially with a set P of elements that contains ($\supset$) three special kinds of subset called *points*, *lines* and *planes*. A plane contains lines and points, a line contains points, and a point is a subset that contains only one element. The set P is a *three-dimensional projective space* if it satisfies the following axioms.

1. Any two distinct points are contained in a unique line.
2. In any plane, any two distinct lines contain a unique common point.
3. Three points that do not lie on one line are contained in a unique plane.
4. Three planes that do not contain a common line contain a unique common point.

Though the other axioms here are true in Euclidean space, and may be regarded as 'self-evident' (consistent with common experience of the usual meaning of 'points', 'lines' and 'planes'), Axioms **2** and **4** seem at first counter-intuitive (we *know* that parallel lines and parallel planes exist in the 'real world'). Their beauty and relevance will emerge as we proceed.

It may seem at first that no theorems of any interest could be expected to arise logically from these apparently trivial axioms. It turns out to be otherwise—a profusion of surprising and beautifully elegant results follow logically.

It will be convenient to employ the following notation: if A and B are two of the special subsets, we introduce the notation AB to denote the smallest special subset that contains both A and B, and $A \cdot B$ to denote the largest special subset that contains both A and B.

Fig. 1.1 *The Annunciation* by Fra Carnevale (15th century)

1.3 The Art of Perspective

Projective geometry has its origins in problems faced by artists and architects. A familiar fact about the way we perceive the world visually is that things look smaller the further away they are, and (consequently) parallel lines do not always *look* parallel. This is so familiar that we take it for granted. The painters and architects of the Italian renaissance did not (Fig. 1.1). If you want a drawing or a painting—particularly of a scene containing buildings—to 'look right' when viewed from the correct viewing position, then a *precise* understanding is needed of the geometrical principles involved when three-dimensional objects are to be represented on a flat surface. Though the problem is easily stated, the technical details involved in its solution are not trivial. They were first thoroughly understood by the 15th century architect Bramante, and his contemporary, the mathematician Alberti.

The problem of course concerns the space in which we live—three-dimensional Euclidean space—so the 'problem of perspective drawing' is a problem in *Euclidean* geometry. In Fig. 1.2 the 'picture plane' p is the flat surface on which a three-dimensional scene is to be depicted and the 'viewing point' E is the intended position of the eye of the beholder. If the depiction has been carried out correctly,

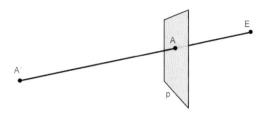

Fig. 1.2 Picture plane p and viewing point E

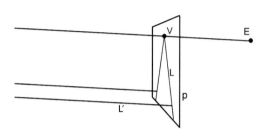

Fig. 1.3 Construction of the vanishing point V for a family of parallel lines

Fig. 1.4 *A Man Drawing a Lute* by Albrecht Dürer

then we can imagine p and E located so that each point A′ of the scene is represented by the point A of intersection of the line A′E and the plane p.

In particular, a straight line L′ in the scene is represented in the picture by L, the line of intersection of p with the plane L′E. It is then obvious that parallel lines that are not parallel to p are represented by lines intersecting at a *vanishing point* V, the intersection of p by a line through E parallel to the given lines (Fig. 1.3).

A rather laborious way of producing a 'perspective' drawing that is correct in this sense, without knowing any of the theory, is shown in a curious device invented by the 16th century artist Dürer (Fig. 1.4). (You would need a lot of string if it were a building instead of a lute...)

Fig. 1.5 Desargues' theorem

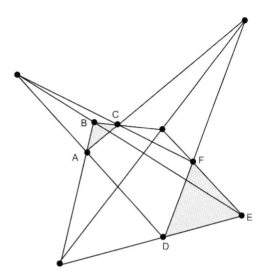

1.4 Desargues' Theorem

Girard Desargues was an architect as well as a mathematician. His knowledge of the theory of perspective drawing led him to develop the conceptual scheme now known as Projective Geometry. The theorem he discovered is a purely *projective* theorem. It is simply a statement about points and lines; the Euclidean concepts of lengths, angles and parallelism are absent.

We call three points *collinear* if they all lie on one line, three lines *concurrent* if they pass through one point. To simplify the statement of Desargues' theorem we shall say that two triangles ABC and DEF are *perspective from a point* if the lines AD, BE and CF joining pairs of corresponding vertices are concurrent. Their common point is called the *vertex* of the perspectivity. We shall say that two triangles ABC and DEF are *perspective from a line* if the three intersections of corresponding pairs of edges, BC · EF, CA · FD and AB · DE, are collinear. The line on which they lie is called the *axis* of the perspectivity. Then the theorem of Desargues states that

> *If two triangles are perspective from a vertex then they are perspective from an axis.*

This is illustrated in Fig. 1.5.

The converse of Desargues' theorem is, of course:

> *If two triangles are perspective from an axis then they are perspective from a vertex.*

If the two triangles are *coplanar* (lie in the same plane) then each of these two statements is an immediate consequence of the other—because Axioms **1** and **2** are interchanged by replacing the words 'point' and 'line', and interchanging the concepts of 'is contained in' (⊂) and 'contains' (⊃). This is the *principle of duality* for plane projective geometry. Any projective statement about points and lines in a

Fig. 1.6 Desargues' theorem in three dimensions

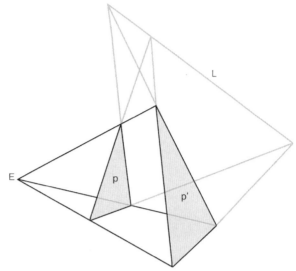

plane has an associated *dual* statement, obtained from it by interchanging the words 'point' and 'line', and interchanging 'lie on' and 'pass through'. Once a theorem is proved the dual theorem must automatically be true and need not be proved separately. In the case of Desargues' theorem in a plane the dual of the theorem is the converse of the theorem. Similarly, the set of axioms for three-dimensional projective geometry is unchanged by the interchange of 'points' and 'planes'. *Duality* is a central concept throughout projective geometry, in any number of dimensions.

A curious characteristic of Desargues' theorem for *coplanar* triangles is that, although the statement of the theorem involves only the concepts of two-dimensional (plane) projective geometry, it cannot be proved only from the axioms for plane projective geometry—Axioms **1**, **2** and **3**. The existence of the third dimension is needed. (It is in fact possible to construct peculiar mathematical structures, called 'non-Desarguesian planes', that contain entities called 'points' and 'lines' satisfying Axioms **1**, **2** and **3**, in which the theorem is *not* true.) But if the two triangles perspective from a point are *not in the same plane* the proof of Desargues' theorem is very simple—almost 'self-evident'.

In the three-dimensional case (Fig. 1.6) two triangles perspective from a point E are in *two different planes*, p and p′, which intersect in a line L (Axiom **4**). (In terms of the theory of perspective drawing outlined above, we can imagine the triangle in p′ as the 'scene' to be depicted, p as the 'picture plane' and E as the 'viewing point'.) Now, two 'corresponding edges' of the two triangles lie in a plane through E, so they intersect (Axiom **2**). But one of these edges is in p and the other is in p′, so their intersection is on L. Similarly, the other two pairs of corresponding edges intersect on L. Which proves the three-dimensional Desargues' theorem.

Of course, Fig. 1.6, which illustrates this, is a *picture* of the three-dimensional figure, and is itself a plane Desargues' configuration. This in mind, we can proceed to employ the three-dimensional theorem to construct a proof of the planar theorem.

1.5 The Complete Quadrilateral

Fig. 1.7 Desargues' configuration labelled by duads

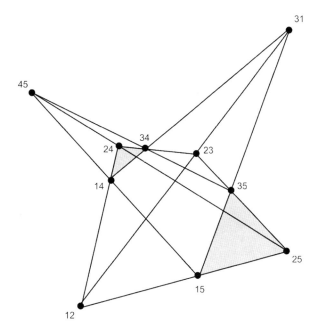

In Fig. 1.7 the ten points of the planar Desargues configuration have been relabeled by duads from the set of symbols **12345**. Observe that the two triangles **14 24 34** and **15 25 35** are perspective from **45**. Erect a line through **45**, not in the plane, and two points A and B on that line (Fig. 1.8). We now have two *non-coplanar* triangles **14** A **34** and **15** B **35**, perspective from the point **45** and therefore perspective from a line L (shown in Fig. 1.6). The lines AB and **24 25** intersect and therefore are coplanar, and so **24** A and **25** B are coplanar and therefore intersect—at some point E. Now look at the two non-coplanar triangles ABE and **14 15 12**. They are perspective from the line **245** and are therefore perspective from a point. That is, E **12**, A **14** and B **15** are concurrent—at a point on L. Thus **12** E intersects L. Similarly, **23** E intersects L and **31** E intersects L. This establishes that the three points **12 23 31** are images, under projection from E, of three points on L. So **23 31 12** are collinear—the *coplanar* triangles **14 24 34** and **15 25 35** are perspective from a line **123**.

1.5 The Complete Quadrilateral

Four lines in a plane have six points of intersection. This is shown in black in Fig. 1.9. This configuration of points and lines is called a *complete quadrilateral*. It looks trivial but is in fact an important figure in plane projective geometry. Incidentally, the six points can be labeled by duads of the four symbols **1234**, just as we labeled the points of Desargues' configuration (Fig. 1.7) with duads from **12345**. The points **12**, **23** and **21** lie on the line **123** (which could be more simply called

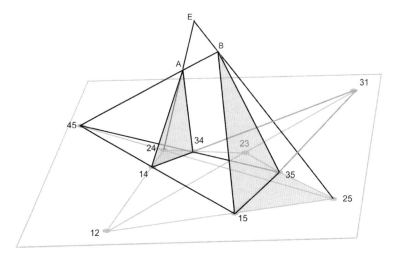

Fig. 1.8 The 3-dimensional Desargues' configuration

Fig. 1.9 A complete quadrilateral

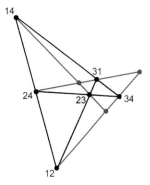

'the line **4**'), and so on. An obvious consequence of this notation is that Desargues' configuration contains five complete quadrilaterals.

Notice that there are three point pairs that do not lie on any of the four lines—the intersections of the 'diagonals' of the quadrilateral. Drawing the lines through them, we get a triangle. This is the *diagonal point triangle* of the quadrilateral. There are now four points on each side of this triangle—two vertices of the diagonal point triangle and two vertices of the quadrilateral. These two point pairs are in a very special relationship. We have a *harmonic range* of four collinear points.

A fundamental theorem in projective geometry states that if a pair of points A and B, and another point C on the line AB, are given, then a fourth point D can be determined uniquely. It is called the *harmonic conjugate* of C with respect to AB. Figure 1.10 shows the method of finding it by constructing a complete quadrilateral. Given A, B and C, choose an arbitrary point F not in the line ABC and an arbitrary point H on CF. AH intersects BF at I and BH intersects AF at G. D is then the

1.5 The Complete Quadrilateral

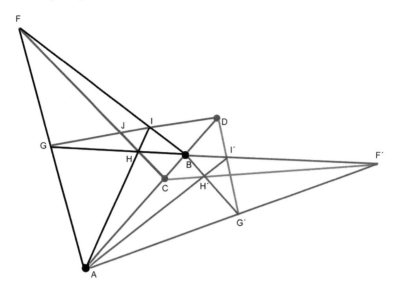

Fig. 1.10 The harmonic construction

intersection of AB and GI. The theorem says, in effect, that it does not matter which points are chosen for F and H, we will always arrive at the *same* point D.

This can be proved from (the three-dimensional) Desargues' theorem. FGHI and F'G'H'I' are two different quadrilaterals used to find D. We see that the triangles FGH and F'G'H' are perspective from the line ABC, so they are perspective from a point E (not shown). The triangles FIH and F'I'H' are also perspective from a point. It is the *same* point E (the intersection of HH' and FF'). So the quadrilaterals FGHI and F'G'H'I' are perspective from E. The triangles GHI and G'H'I' are perspective from E and hence are perspective from a line. In other words, it follows that GI and G'I' intersect at a point on AB. This is the unique point D.

It is instructive to think of this construction in terms of the traditional theory of perspective drawing. Imagine a piece of paper on which Fig. 1.10 is drawn. Fold it along the line AB, so that FGHI and F'G'H'I' are in different planes. We can think of one of the figures FGHI as the 'scene' and F'G'H'I' as its representation in the picture plane (or, of course, vice versa). E is then the viewing point. Equivalently, think of F'G'H'I' as the shadow of FGHI cast by a point light source at E.

Let us abridge the statement that D is the harmonic conjugate of C with respect to the point pair AB', writing it simply as {AB, CD}. It is obvious that this is the same thing as {BA, CD} or {AB, DC}. Less obvious is that it implies {CD, AB}. Figure 1.11 shows why. The black quadrilateral FGHI is constructed as described above for obtaining D, the harmonic conjugate of C with respect to the point pair AB. Its diagonal triangle is CDJ. We have (AB, CD). The quadrilateral KLMN inscribed in FGHI is constructed as follows. L and N are the intersections of AJ with GH and IF, respectively, and M and K are the intersections of BJ with HI and FG. We see that the two triangles LKG and NMI are perspective from J. So they are perspective from a

Fig. 1.11 Proof that
{AB, CD} implies {CD, AB}

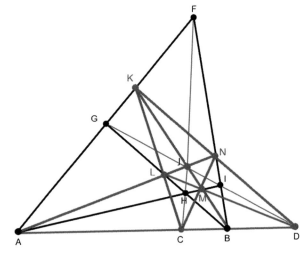

line: $KG \cdot MI = A$, $GL \cdot IN = B$ and $LK \cdot MN = C'$ are collinear. Similarly, NKF and LMH are perspective from J, so $KF \cdot MH = A$, $FM \cdot HL = B$ and $NK \cdot LM = D'$ are collinear. Now note that LKG and MNI are perspective from D'. They are therefore perspective from a line: $KG \cdot NI = F$, $GL \cdot IM = H$ and $LK \cdot MN = C'$ are collinear. Therefore C' is on FH as well as on A. So $C' = C$. Similarly, $D' = D$. The quadrilateral KLMN then indicates that B is the harmonic conjugate of A with respect to the point pair CD. That is, {AB, CD} implies {CD, AB}.

1.6 Affine Geometry

Those theorems and constructions of Euclidean geometry that are devoid of metrical concepts, but in which the concept of *parallelism* is retained, constitute *Affine geometry*. Euclid's postulates I and II and Playfair's axiom V' can be taken as its axioms.

Observe first that it follows from Euclid's postulates I and II, together with Playfair's axiom V', that if A, B and C are three lines, and if A ∥ B and B ∥ C, then C ∥ A (parallelism is transitive). Also, trivially, if A ∥ B then B ∥ A (parallelism is symmetric). We may conveniently extend the concept of parallelism by agreeing to the convention that a line is parallel to itself, A ∥ A (parallelism is reflexive). Thus *parallelism is an equivalence relation*. The equivalence classes are *families of parallel lines*. In an affine plane, or an affine space of three or more dimensions, any given line belongs to a family of parallel lines. Exactly one member of the family passes through any chosen point.

A simple way of getting an affine theorem from a projective theorem is suggested by replacing the Euclidean 'parallel lines never meet' by the more casual and informal 'parallel lines meet at infinity'—we know intuitively what is meant. So let us single out a line in a projective plane and call it 'the line at infinity'. The points on

1.7 The Theorem of Pappus

Fig. 1.12 An affine specialization of Desargues' theorem

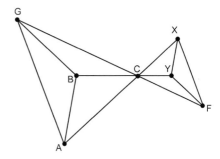

this special line are called 'points at infinity'. Then parallelism in the plane is introduced into the projective plane, without discarding the projective Axiom **2**. Lines are then said to be 'parallel' if their point of intersection is 'at infinity'. In this way, we have constructed a model of an affine plane in a projective plane. This model satisfies the axioms of affine geometry. In this approach, all the lines of a family of parallel lines pass through the same 'point at infinity'—thus a 'point at infinity' corresponds to the *orientation* of a family of parallel lines in an affine (or a Euclidean) plane. Similarly, a three-dimensional projective space is specialized to a three-dimensional affine space by singling out a plane and calling it 'the plane at infinity'. All the points and lines in that plane are said to be 'at infinity'. Lines are 'parallel' if their intersection is a point at infinity and planes are 'parallel' if their line of intersection is at infinity.

We can appreciate the greater logical simplicity that has come from throwing away the Euclidean concept of parallel lines in favor of the projective Axiom **2**, if we think about what happens to Desargues' theorem in Euclidean geometry or Affine geometry. Then one or more of the ten points involved in the construction indicated by Fig. 1.5 may be 'at infinity'—that is, there may be one or more sets of two or three *parallel* lines. We would need to consider a quite large number of separate special cases, corresponding to different theorems, instead of a single simply stated theorem. Each of these is a Euclidean theorem that makes no appeal to *metrical* concepts (angles, lengths, areas *et cetera*), but may involve the concept of parallel lines. These statements belong to Affine geometry.

Figure 1.12 illustrates just one of the affine special cases of the planar Desargue's theorem (corresponding to choosing DE of Fig. 1.5 as the 'line at infinity'):

> Let ABC be a triangle and G a point not on any of its sides. Let F be a point on GC. Let X be the intersection of AC with the line through F parallel to AG. Let Y be the intersection of BC with the line through F parallel to BG. Then XY ∥ AB.

1.7 The Theorem of Pappus

Pappus of Alexandria lived in the fourth century, when geometers were preoccupied with metrical concepts, following the Euclidean tradition. His theorem has quite a

Fig. 1.13 Pappus's affine theorem

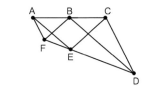

Fig. 1.14 The projective Pappus theorem

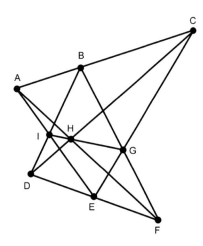

different flavor from other geometrical theorems known at that time. The statement of the theorem he proved is:

> *If three distinct points* A, B *and* C *are collinear and three different distinct points* D, E *and* F *in the same plane are collinear, and if* AE ∥ BD *and* CD ∥ AF, *then* BF ∥ CE (Fig. 1.13).

Though this is a purely *affine* statement, it is not a consequence of the affine axioms I, II and V′ alone. 'Non-Pappian' affine planes can be constructed in which these three axioms are satisfied, but in which the theorem is not true. Its proof requires further assumptions from *Euclidean* geometry, such as the properties of 'similar triangles'.

The theorem now known as Pappus's theorem is the projective generalization:

> *If three distinct points* A, B *and* C *are collinear and three different distinct points* D, E *and* F *in the same plane are collinear, then the three points of intersection* G = BF · CE, H = CD · AF *and* I = AE · BD *are also collinear* (Fig. 1.14).

The theorem that Pappus discovered is the special case in which the line GHI is the 'line at infinity'.

The theorem is sometimes called 'Pappus's hexagon theorem' because it can be restated as: *if the vertices of a planar hexagon lie alternately on two lines, then the three intersections of opposite edges are collinear*. The hexagon AFBDCE in the figure, for example, illustrates this property.

1.7 The Theorem of Pappus

Fig. 1.15 Pappus implies Desargues

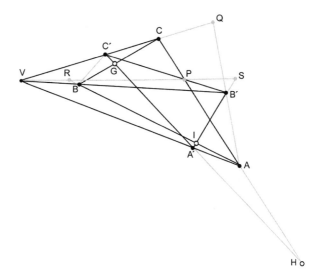

Recall that, though the statement asserted by the planar version of Desargues' theorem involves only *projective* concepts (incidence properties of points and lines), it is not a logical consequence of Axioms **1**, **2** and **3** of the projective plane. Some further assumption was needed, and we showed that Axiom **4** sufficed—the planar Desargues theorem was proved by assuming that the plane in question was a plane in a three-dimensional projective space. Similarly, although the statement asserted by Pappus's original theorem is a purely *affine* statement, the theorem is *not necessarily true* without further assumptions. It can be proved in a *Euclidean* plane by making use of the properties of 'similar triangles'—and these are *not* affine properties.

Similarly, though the statement asserted by the *projective* version of the Pappus theorem involves only projective concepts (incidence properties of points and lines), it is not a logical consequence of Axioms **1**, **2** and **3** of the projective plane. There exist mathematical curiosities, called 'non-Pappian planes', that satisfy these axioms, but in which Pappus's theorem is not true. Nor does appeal to higher dimensions work, as it did for the planar Desargues' theorem. Thus the Pappus theorem is not a *projective* theorem at all. It is a *Euclidean* theorem. *The simplest way out of this situation is to accept it as an additional axiom of projective geometry.* Further interesting results then follow. For example:

The planar Desargues' theorem is a consequence of Pappus's 'theorem'.

That is, if Pappus's theorem is *assumed* to be true, then, on the basis of Axioms **1**, **2** and **3**, Desargues' theorem is also true. (The converse is not true.) To demonstrate this requires three applications of Pappus's theorem. In Fig. 1.15, the triangle ABC and DEF are perspective from the point V. We are required to prove, assuming Pappus's theorem, that G, H and I are collinear. Construct the points $P = B'C' \cdot AC$, $Q = AB' \cdot VC$ and $R = AB \cdot VP$. $P = B'C' \cdot AC$, $Q = AB' \cdot VC$ and $R = AB \cdot VP$. Then the vertices of the hexagon ABCVPB' lie alternately on the lines ACP and

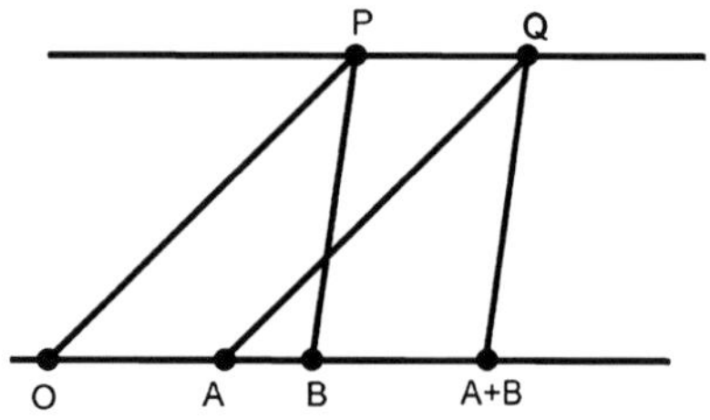

Fig. 1.16 Addition on an affine line

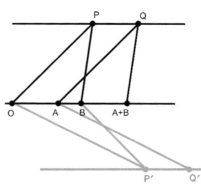

Fig. 1.17 Addition on an affine line, demonstrating arbitrariness of the supplementary point P

BVB′. Therefore, according to the Pappus theorem, G, Q and R are collinear. Construct S = A′B′ · VP. The vertices of the hexagon B′APVC′A′ lie alternately on the lines VAA′ and B′C′P. Therefore H, S and Q are collinear. Finally, the vertices of the hexagon ARQSB′P lie alternately on the lines AQB′ and SPR.

Therefore G, H and I are collinear.

1.8 Affine Coordinates

It is a remarkable fact that

> *If Pappus's affine 'theorem' (and consequently Desargues' theorem) is valid in an affine plane, then the points on a line in that plane can be added and multiplied in a way that satisfies the laws of arithmetic.*

Figure 1.16 shows how, when a point O is selected as an origin on a line AB in an affine space, the *addition* of two points A and B can be defined. Choose any point P not on AB, Let Q be the intersection of the line through A parallel to OP, and the line through P parallel to AB. The line through Q parallel to PB then intersects AB at a point which we call A + B. This point A + B is independent of the choice of the point P.

In Fig. 1.17, an alternative choice P′ has been made. The two triangles OPP′ and AQQ′ are perspective from a 'point at infinity', therefore, according to Desargues' theorem, they are perspective from an axis. But, by construction, OP ∥ AQ and OP′ ∥ AQ′, the axis is therefore the 'line at infinity', so we have PP′ ∥ QQ′. Then, similarly,

1.8 Affine Coordinates

Fig. 1.18 Commutativity of addition

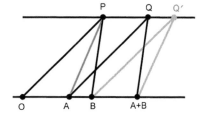

Fig. 1.19 Additive inverse

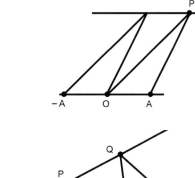

Fig. 1.20 Multiplication on an affine line

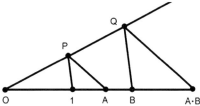

the perspectivity of the two triangles PP′B and QQ′(A+B) implies that Q′(A+B) ∥ P′B. The construction starting from P′ therefore leads to the same *unique* point A + B.

Figure 1.18 demonstrates the commutativity of this definition of addition. The construction for locating A + B and the construction for locating B + A have been superimposed. The hexagon APBQ′(A+B)Q has alternate vertices PB ∥ Q(A+B) and AQ ∥ BQ′. That is, the intersection of PB with Q(A + B) and the intersection of AQ with BQ′ are on the line at infinity. Therefore, *according to Pappus's affine theorem*, the intersection of PA with Q′(A + B) is also on the 'line at infinity': PA ∥ Q′(A + B). This establishes that B + A coincides with A + B.

That O is the identity for addition (A + O = A) is fairly obvious, as is the existence of a unique additive inverse −A for every point A (A + (−A) = O). The construction is illustrated in Fig. 1.19.

The *multiplication* of two points is shown in Fig. 1.20. An origin O and a unit point 1 are chosen on an affine line.

As before, P is an arbitrary point. Q is the intersection of OP and the line through B parallel to 1P. Then A · B is the intersection of O1 with the line through Q parallel to PA. The independence of the position of A · B on the choice of P follows from two applications of Desargues' theorem (as we showed in the case of A + B). The commutativity of multiplication, like the commutativity of addition, is a consequence of Pappus's theorem (Fig. 1.18). Again, by superimposing the constructions

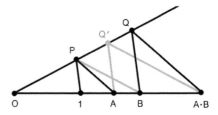

Fig. 1.21 Commutativity of multiplication

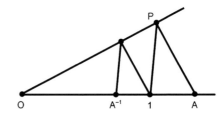

Fig. 1.22 Multiplicative inverse

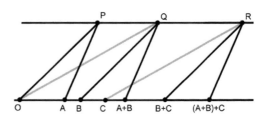

Fig. 1.23 Associativity of addition

for $A \cdot B$ and $B \cdot A$ (Fig. 1.21), we see a Pappus hexagon, and the Pappus theorem guarantees that $Q'(A \cdot B)$ is parallel to PB. It is a simple matter to deduce that the point A^{-1} constructed according to Fig. 1.22 satisfies $A \cdot A^{-1} = 1$, and trivial to see that $A \cdot O = O$.

The associative laws for addition and multiplication, and the distributive law, remain to be demonstrated. In Fig. 1.23, O, A, B and C are four collinear points. The point $A + B$ is constructed with the aid of an auxiliary point P. This construction locates an intermediate point Q on the line through P parallel to the line AB. Then $B + C$ is located using the auxiliary point Q. The construction locates an intermediate point R on the line PQ. This point R is used as the auxiliary point for finding $(A + B) + C$. The final step involves the line through R parallel to $Q(A + B)$. The analogous procedure for locating $A + (B + C)$ involves the line through R parallel to PA. But $PA \parallel Q(A + B)$. Hence $(A + B) + C = A + (B + C)$. An exactly analogous argument applied to Fig. 1.24 establishes that multiplication is also associative, $(A \cdot B) \cdot C = A \cdot (B \cdot C)$.

Finally, the distributive law is established with the visual aid of Fig. 1.25. Using the auxiliary point P, construct $A \cdot B$ and $A \cdot C$. These constructions identify two points Q and R on OP. Observe how the auxiliary point Q has been used to construct the points $B + C$ and $A \cdot B$ and $A \cdot C$. The triangles STU and RC $(A \cdot C)$ are perspective from a 'point at infinity'—because $RS \parallel (A \cdot C)U \parallel CT$. Therefore, by Desargues', they are perspective from an axis. This axis is the 'line at infinity'

1.9 Configurations

Fig. 1.24 Associativity of multiplication

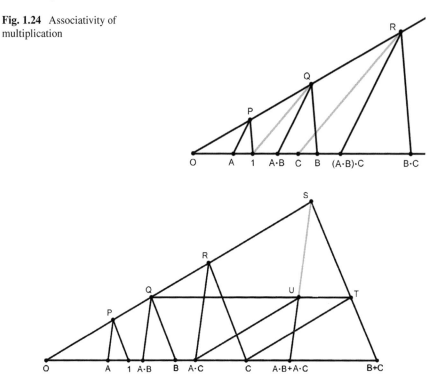

Fig. 1.25 The distributive law

because ST ∥ RC and TU ∥ C(AC). Therefore SU ∥ R(A · C). When A(B · C) is constructed, the final step requires the line through S parallel to PA. but PA ∥ R(A · C), so this is the line SU. Hence $A \cdot (B + C) = A \cdot B + A \cdot C$.

It is now clear that in any Desarguesian affine plane in which Pappus's theorem is valid, we can choose an origin O, two lines through O and a 'unit point' on each line. The lines are then coordinate axes. Every point in the plane can be referred to uniquely by a pair of points X and Y on the axes. The point in question then has coordinates (X, Y) (Fig. 1.26). Note that there is no requirement that X and Y be real numbers. They can be any entities that satisfy the laws of arithmetic for addition and multiplication. The concept is of course generalizable to affine spaces of three or more dimensions.

1.9 Configurations

In this section we digress a little to introduce some ideas that we shall need later. In a projective plane, a set of p 'points' and l 'lines', with m lines through each point and n points on each line is often referred to as a *configuration* $(p_m l_n)$ of points and lines. Then, necessarily, $pm = ln$. A complete quadrilateral, for example, is a

Fig. 1.26 Coordinate axes in an affine plane

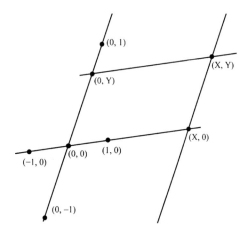

Fig. 1.27 Fano's (7_3)

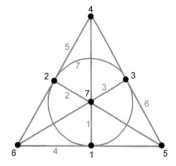

$(6_2 4_3)$. Its dual is a $(4_3 6_2)$: four points, no three of which are collinear, and the six lines through pairs of them. This is called a *complete quadrangle*. A symbol $(p_m p_m)$ is abridged to (p_m). The planar Desargues' configuration (Fig. 1.7), for example, is a (10_3). The three-dimensional Desargues' configuration is a $(10_3 5_6)$ of points and planes.

Configurations can be considered in a purely abstract way, without reference to their representation in terms of geometrical figures. For example, in Fig. 1.9 we labeled the six points of a complete quadrilateral with the duads from the set of four symbols **1234** and the lines can correspondingly be labeled **1, 2, 3, 4**. We can simply define *these symbols* to be the 'points' and 'lines' of a $(6_2 4_3)$. We can then consider the *symmetries* of the configuration as the permutations of the symbols that preserve all the incidence relations. In this simple case the group of symmetries obviously consists of all possible permutations of **1234**—a group of order 24.

A more interesting example is Fano's configuration (7_3), consisting of seven 'points' **1234567** and seven 'lines' *1234567*. Such a configuration is impossible if by 'lines' we mean straight lines of Euclidean geometry. Hence one of the 'lines' is represented by a circle in Fig. 1.27. The configuration is defined by an *incidence table* specifying, for each point, the three lines through it. There is essentially only

1.9 Configurations

Fig. 1.28 Levi graph for (7_3)

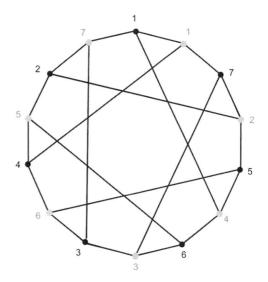

one way this can be done (apart from trivial relabeling) so that every line goes through just three of the points. This table is on the left, below. On the right is the table constructed from it, listing for each line the three points on it:

1	147	1	147
2	257	2	257
3	367	3	367
4	156	4	156
5	264	5	264
6	345	6	345
7	123	7	123

The fact that these two tables are identical apart from interchange of points and lines demonstrates the *self-duality* of the configuration.

Some of the incidence-preserving symmetries are readily apparent from Fig. 1.27—a group D_3 of order 6 generated by the three permutations **(23)(56)**(*23*)(*56*), **(31)(64)** (*31*)(*64*) and **(12)(45)**(*12*)(*45*).

An interesting way of visualizing the incidence table of a (p_m) is by its *Levi graph*, a configuration with $2p$ vertices and $2pm$ lines. The vertices of the Levi graph correspond to the points *and* the lines of the (p_m) and the incidences of the (p_m) are indicated by the lines of the graph. Figure 1.28 is a Levi graph for Fano's (7_3). Immediately apparent from the graph is the fact that all vertices of the (7_3) are equivalent, as are all its edges. The configuration is *regular* in the sense that any point can be mapped to any other by a symmetry transformation—a fact that is not at all apparent in Fig. 1.27.

The duality **(1***1***)(2***2***)(3***3***)(4***4***)(5***5***)(6***6***)(7***7***)**, already noticed, corresponds to one of the seven reflection symmetries of Fig. 1.33. The cyclic subgroup of order 7 gener-

Fig. 1.29 Levi graph for (7₃) on a lattice

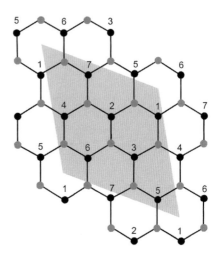

ated by the permutation (**1756342**)(*2436571*) corresponds to the rotational symmetries of the Levi graph. The outer 14-gon of Fig. 1.28 corresponds to a *Hamiltonian circuit* of the (7₃)—a path in the configuration that traverses every vertex and every edge just once and returns to the starting point. The (7₃) can be described as a *self-inscribed and circumscribed heptagon*: every edge of the heptagon **1756342** contains just one vertex (in addition of course to the vertex pair that determines the edge).

The Levi graph can be nicely represented on a lattice, as in Fig. 1.29, with seven grey and seven black points in a unit cell (indicated by the grey rhombus). If we imagine opposite sides of this rhombus identified, we have the graph represented on a torus.

The whole symmetry group of the (7₃) is a group of order 168. To see why, observe that any complete quadrilateral in Fig. 1.27 determines the whole configuration (its diagonal triangle collapses to the single remaining point!). Since the configuration is self-dual we can say, equivalently, that any complete *quadrangle* determines the whole configuration. So if we count the number of ways of choosing four of the points, *no three collinear*, that will be the number of ways of permuting **1234567** that preserve all the incidences. We can choose the first point in seven ways, the second in six ways, the third in four ways. There is then only one point left that is not collinear with any pair of these three. $7 \times 6 \times 4 = 168$.

The Pappus configuration (9₃) illustrated in Fig. 1.30 is not the only kind of (9₃) but it is the most interesting. It is, of course, the figure described by Pappus's hexagon theorem, Observe that *any* two lines of the configuration that have no common point determine a hexagon.

The order of the symmetry group of the (9₃) is 108. Figure 1.31 shows how the Pappus (9₃) can be represented on a lattice or a torus.

Figure 1.32 is a Levi graph for the Pappus (9₃). The self-duality of the theorem is revealed by the mirror symmetries of this version of the graph. The outer 18-gon represents a Hamiltonian circuit on the (9₃): a self-inscribed nonagon **146329758**.

1.9 Configurations

Fig. 1.30 The Pappus (9_3)

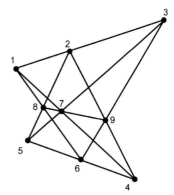

Fig. 1.31 The Pappus (9_3) on a lattice

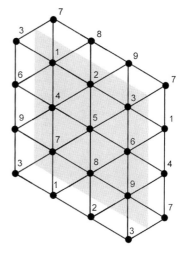

Fig. 1.32 Levi graph for the Pappus configuration

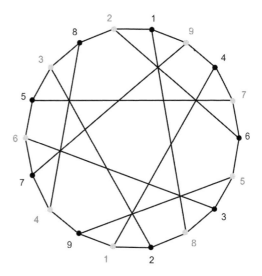

Fig. 1.33 Levi graph for the Pappus (9₃) on a lattice

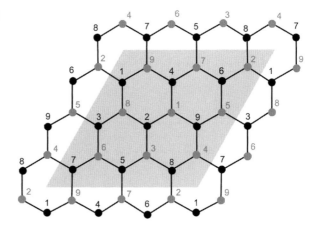

Fig. 1.34 Levi graph for Desargues' (10₃)

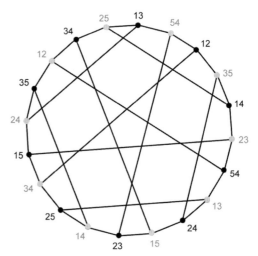

Another is **149652378**. The (9₃) can also be described as three triangles A, B and C, with A inscribed in B, B inscribed in C, and C inscribed in A—for example **136**, **982** and **547**.

A delightfully simple way of representing the incidences of the Pappus (9₃) is by indicating its Levi graph on a lattice (Fig. 1.33). The grey rhombus is a unit cell.

In Fig. 1.7 the points of Desargues' (10₃) were labeled with the ten duads of symbols formed from the symbol set **12345**, so that, for example, the three points **23**, **31** and **12** lie on a line which may be called the line **123** or simply the line **45**. This labeling system has the obvious property that any permutation of the points and lines given by a permutation **12345** along with the corresponding permutation *12345* will leave all the incidence properties of the Desargues' (10₃) configuration intact. The *symmetry group* of the configuration is therefore just the group of permutations of five objects, a group of order 5! = 120. The configuration is *regular*. In addition, lines and points can be interchanged (interchange of grey and black in our

1.9 Configurations

Fig. 1.35 An alternative version of the Levi graph for Desargues' (10_3)

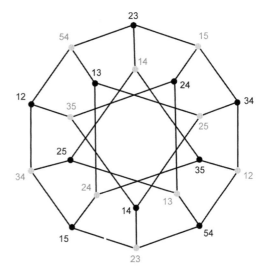

Fig. 1.36 A self-dual (10_3)

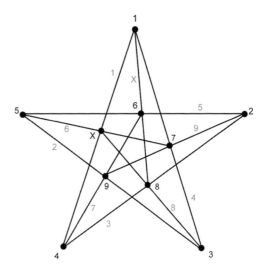

labeling scheme), and again all the incidences are preserved. This is the *self-duality* of the configuration. Figures 1.34 and 1.35 are two (topologically equivalent) representations of the Levi graph of the Desargues' (10_3). The first of these graphs demonstrates the existence of a Hamiltonian circuit on the (10_3)—the configuration is a self-inscribed decagon. The second version reveals that the Desargues' (10_3) can also be regarded as *a pair of mutually inscribed pentagons*!

The Desargues configuration is not the only kind of (10_3). There are no less than ten different (10_3)s. Figure 1.36 shows one. Its points and lines have been labeled in a way that reveals that the figure is *self-dual*—**5** lies on **2**, **5** and **6**, and also **5** contains **2**, **5** and **6**, and so on. Its Levi graph is shown in Fig. 1.37.

Fig. 1.37 Levi graph for Fig. 1.36

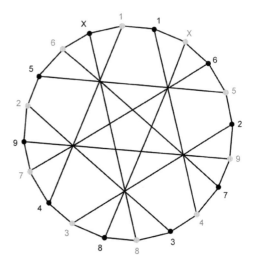

We shall later meet many more configurations, in three and more dimensions. We now turn to the method of coordinates in the projective plane, which is somewhat different from the use of coordinates in Euclidean geometry, and rather more elegant.

1.10 Axioms for N-Dimensional Projective Space

In this section we suggest one possible approach to the setting up of axioms for N-dimensional projective space. An N-dimensional projective space F is a collection of things (including F itself) called *subspaces*, denoted by P, Q, etc., satisfying the following properties:

Given P and Q belonging to F, there is a unique third subspace belonging to F, called the *intersection* of P and Q, denoted by $P \cdot Q$.

Given P and Q belonging to F, there is a unique third subspace belonging to F, called the *join* of P and Q, denoted by PQ.

A1. Join and intersection are *commutative* and *associative* (PQ = QP and P(QR) = (PQ)R = PQR and, similarly, $P \cdot Q = Q \cdot P$ and $P \cdot (Q \cdot R) = (P \cdot Q) \cdot R = P \cdot Q \cdot R$
A2. $P \cdot (PQ) = P(P \cdot Q) = P$
A3. F is an *identity* for intersection: $F \cdot P = P$ for any P
A4. There is an *identity* for join, called the 'null space', which we shall denote by the symbol E. It is defined by EP = P for any P

We shall take these as *axioms*. From them, it is easy to prove that

The identities E and F are unique.
$P \cdot Q = P$ if and only if $PQ = Q$
$P \cdot E = E$
$PF = F$

$P \cdot P = P$
$PP = P$

We *define* 'contains', denoted by $\supset$, and 'is contained in', denoted by $\subset$ as follows:

$P \subset Q$ (equivalently, $Q \supset P$) means $P \cdot Q = P$ (equivalent to $PQ = Q$).

It is straightforward to show that $\supset$ is *transitive* (if $P \supset Q$ and $Q \supset R$, then $P \supset R$), and so is $\subset$. It is also easily proved that

$P \cdot Q \subset P$
$P \subset PQ$
For any P, $E \subset P \subset F$
If $P \subset Q$ and $Q \subset P$, then $P = Q$
If $P \subset R$ and $Q \subset R$ then $PQ \subset R$
If $R \subset P$ and $R \subset Q$ then $R \subset P \cdot Q$

It is not difficult, and is a useful exercise in logical deduction, to prove all these simple results from Axioms A1 to A4.

Finally, we need to deal with the *dimensionality* of projective spaces. To every P, assign an integer belonging to the set $-1, 0, 1, 2, \ldots, N$, called the *dimensionality* of P, which we denote by D(P). We choose the *axioms of dimensionality* to be

A5. $D(E) = -1$
A6. $D(F) = N$
A7. For every integer n between -1 and N there is a P for which $D(P) = n$
A8. If $P \subset Q$ then $D(P) \leq D(Q)$
A9. $D(P \cdot Q) = D(P)$ implies $P \subset Q$
A10. $D(PQ) = D(P)$ implies $Q \subset P$
A11. $D(P) + D(Q) = D(PQ) + D(P \cdot Q)$

1.11 Duality

Observe that the list of Axioms A1 to A11 remains unaffected if we make the following changes throughout:

- $D(P) \to N - 1 - D(P)$
- $E \leftrightarrow F$
- $P \cdot Q \leftrightarrow PQ$
- $\subset \leftrightarrow \supset$

This is the *principle of duality* for N-dimensional projective geometry. Given any theorem proved from the axioms, a related *dual* theorem can be obtained from it immediately by making these changes. It does not require separate proof.

1.12 Algebra *versus* Axioms

There are two distinct approaches to geometry: the *synthetic* and the *analytic*. The synthetic approach proceeds from a set of *axioms*, and deduces theorems that arise from them by logical deduction. In this purely axiomatic approach to geometry we have simply a set of symbols denoting undefined 'things', and rules for manipulating these symbols and making logical deductions. This was the method adopted in the foregoing discussion. Of course, the undefined things and their relationships must correspond to our intuitive understanding of what we mean by 'points', 'lines', 'planes', etc.—otherwise it would not be 'geometry' at all. The analytic approach treats geometrical problems as *algebraic* problems. The simplest and most well known way of doing this for three-dimensional *Euclidean* geometry proceeds by choosing an origin and a set of three orthogonal axes—the method of Cartesian coordinates, named after its originator René Descartes. This idea provided powerful methods of solving geometrical problems by treating the points as *sets of real numbers* and thereby converting geometrical problems into *algebraic* problems.

Affine geometry can be dealt with similarly by setting up coordinate axes through an origin, but in this case there is no concept of orthogonality and lengths of two line segments are only comparable if they are parallel.

The analytic approach to projective geometry deals with the algebraic properties of 'homogeneous coordinates'. These coordinates are *numbers* which belong to a *number field* F. They may be real numbers, complex numbers, or numbers belonging to some more exotic number field. The N-dimensional projective space of this kind is then called $P(N, F)$.

We now turn attention to analytic methods for dealing with projective geometry.

Chapter 2
The Analytic Approach

Abstract Homogeneous coordinates for a projective plane are introduced and extended to more than two dimensions. This leads to the concept of projective collineations. The theorems of Pappus and Desargues are proved using the method of homogeneous coordinates. The relation between analytic Projective geometry of N-dimensions and vector space theory is established. The concept of projective collineations is introduced and the classification of collineations in the projective plane described. Plücker's line coordinates are then briefly discussed and their N-dimensional generalization to Grassmann algebra is explained.

2.1 Homogeneous Coordinates

The analytic approach to projective geometry is based on a system of *homogeneous coordinates*.

To begin thinking about how coordinates can be assigned in projective geometry, it is illuminating to recall the historical origins of projective geometry in the techniques used by artists and architects for representing three-dimensional objects in a picture.

In Fig. 2.1, we can think of the grey plane p as a picture plane and E as the viewing point. Then any point P′ in space is represented on the plane by the point P where the line EP′ intersects the plane. Take E as the origin and set up three coordinate axes through E. These may be axes for an *affine* coordinate system in three-dimensional space, so they need not be orthogonal and the length units along the three axes—given by the line segments connecting the origin to the points (100), (010) and (001)—need not be related. Hence the positions of these three points along the three axes can be chosen arbitrarily.

It will be convenient to avoid the constant repetition of the phrase 'the point with coordinates $(X\ Y\ Z)$'. We shall often simply say 'the point $(X\ Y\ Z)$'. Indeed, no harm is done by regarding a 'point' as a set of numbers.

For the present, we shall take the coordinates to be *real numbers*. We are dealing with the 'two-dimensional projective space over the reals', P(2, R).

Now if P′ is $(X\ Y\ Z)$, its image P in the plane is $(\lambda X\ \lambda Y\ \lambda Z)$ for some λ. The important thing to realize is that it *does not matter* what λ is. So long as it is not zero it can be chosen *arbitrarily*, and $(\lambda X\ \lambda Y\ \lambda Z)$ still serves to identify the point P

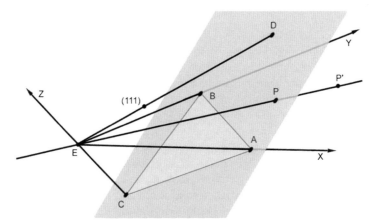

Fig. 2.1 Projection of three-dimensional affine space onto a plane

in the plane. Only the ratios $X : Y : Z$ are significant, which is why the points of the plane p—a *two*-dimensional space—now have *three* 'homogeneous' coordinates.

Notice there are no coordinate *axes* in the projective plane p. Instead, there is a *reference triangle* ABC, given by the intersections of the axes EX, EY and EZ of the affine space with the plane p. The vertices A, B and C of the reference triangle are (100), (010) and (001). The point (111) in the three-dimensional affine space can be chosen arbitrarily by adjusting the scales along the three axes. This does not change the position of the reference triangle ABC, and completes the reference system in the projective plane p by introducing the *unit point* D, with homogeneous coordinates (111). It is obvious from Fig. 2.1 that *any* four points in p, no three of which are collinear, can be chosen to be the reference triangle ABC and the unit point D. That's just a matter of choosing the set of axes through the origin E and the position of the point (111) in the three-dimensional affine space.

The *lines* in the projective plane also have homogeneous coordinates. A plane through E has an equation $lX + mY + nZ = 0$. It intersect the projective plane p in a line L, which acquires the homogeneous coordinates $[l\ m\ n]$. The condition for a point $(X\ Y\ Z)$ and a line $[l\ m\ n]$ in the projective plane to be incident is then simply

$$lX + mY + nZ = 0.$$

Now, if we choose another plane p′ (not through E), with a different position and orientation, any diagram (drawing) in p is projected to a diagram in p′—a distorted version of the drawing in p. But the two diagrams look the same when viewed from E. We regard the two diagrams as *projectively equivalent*. This kind of equivalence has a role similar to *congruence* in Euclidean geometry.

If two points $(X_1\ Y_1\ Z_1)$ and $(X_2\ Y_2\ Z_2)$ are given, the line $[l\ m\ n]$ through them is easily found. The two conditions

$$lX_1 + mY_1 + nZ_1 = 0$$

2.1 Homogeneous Coordinates

$$lX_2 + mY_2 + nZ_2 = 0,$$

are satisfied by

$$[l\ m\ n] = [Y_1Z_2 - Z_1Y_2, Z_1X_2 - X_1Z_2, X_1Y_2 - Y_1X_2].$$

It is convenient to write this as

$$[l\ m\ n] = \begin{vmatrix} X_1 & Y_1 & Z_1 \\ X_2 & Y_2 & Z_2 \end{vmatrix}.$$

The coordinates $[l\ m\ n]$ are then got by 'cross multiplication', as in the cross product of two vectors, or, equivalently, they are the coefficients of X, Y and Z when the determinant

$$\begin{vmatrix} X & Y & Z \\ X_1 & Y_1 & Z_1 \\ X_2 & Y_2 & Z_3 \end{vmatrix}$$

is expanded:

$$[l\ m\ n] = \left[\begin{vmatrix} Y_1 & Z_1 \\ Y_2 & Z_2 \end{vmatrix}, \begin{vmatrix} Z_1 & X_1 \\ Z_2 & X_2 \end{vmatrix}, \begin{vmatrix} X_1 & Y_1 \\ X_2 & Y_2 \end{vmatrix} \right].$$

Similarly (and *dually*), the point of intersection of two lines $[l_1\ m_1\ n_1]$ and $[l_2\ m_2\ n_2]$ is given by

$$(X\ Y\ Z) = \begin{vmatrix} l_1 & m_1 & n_1 \\ l_2 & m_2 & n_2 \end{vmatrix} = (m_1n_2 - n_1m_2, n_1l_2 - l_1n_2, l_1m_2 - m_1l_2).$$

The sides of the reference triangle, of course, are the lines [100], [010] and [001].

There is one plane through E that does not intersect the plane p. It is the plane through E *parallel* to p. In a manner of speaking we may say that this plane corresponds to the 'line at infinity' in p. This fictitious line has coordinates $[l\ m\ n]$ just like any other line, and under the projection from p to p' it can correspond to an actual line. So that, to maintain the concept of *projective equivalence*, we need to regard the 'line at infinity' as no different from any other line. The geometry in any plane p not through E is then a true *projective geometry*—satisfying Axioms **1**, **2** and **3**.

Three points $(X_1\ Y_1\ Z_1)$, $(X_2\ Y_2\ Z_2)$ and $(X_3\ Y_3\ Z_3)$ are collinear if and only if

$$lX_1 + mY_1 + nZ_1 = 0$$
$$lX_2 + mY_2 + nZ_2 = 0$$
$$lX_3 + mY_3 + nZ_3 = 0$$

for some triple $[l\ m\ n]$. This condition is the same as the vanishing of the determinant

$$\begin{vmatrix} X_1 & Y_1 & Z_1 \\ X_2 & Y_2 & Z_2 \\ X_3 & Y_3 & Z_3 \end{vmatrix}.$$

That is to say, the rows of this matrix are linearly dependent; the coordinates of the three points satisfy a relation of the form

$$\lambda(X_1\ Y_1\ Z_1) + \mu(X_2\ Y_2\ Z_2) + \nu(X_3\ Y_3\ Z_3) = 0.$$

The *dual* of all this is:

Three lines $[l_1\ m_1\ n_1]$, $[l_2\ m_2\ n_2]$ and $[l_3\ m_3\ n_3]$ are concurrent if and only if

$$Xl_1 + Ym_1 + Zn_1 = 0$$
$$Xl_2 + Ym_2 + Zn_2 = 0$$
$$Xl_3 + Ym_3 + Zn_3 = 0$$

for some triple $(X\ Y\ Z)$. This condition is the same as the vanishing of the determinant

$$\begin{vmatrix} l_1 & m_1 & n_1 \\ l_2 & m_2 & n_2 \\ l_3 & m_3 & n_3 \end{vmatrix}.$$

That is to say, the rows of this matrix are linearly dependent; the coordinates of three concurrent lines satisfy a relation of the form

$$\lambda[l_1\ m_1\ n_1] + \mu[l_2\ m_2\ n_2] + \nu[l_3\ m_3\ n_3] = 0.$$

2.2 More than Two Dimensions

All this is readily generalizable, in an obvious way, to N-dimensional projective spaces, in which the points and *hyperplanes* (($N-1$)-dimensional subspaces) have $N+1$ homogeneous coordinates. In projective 3-space, for example, a point has four homogeneous coordinates $(X\ Y\ Z\ W)$ and a *plane* has four homogeneous coordinates $[l\ m\ n\ k]$. Any four non-coplanar points can be chosen as the vertices (1000), (0100), (0010) and (0001) of the *reference tetrahedron* and the freedom remains to designate any other point not on a face of the tetrahedron as the unit point (1111). A point $(X\ Y\ Z\ W)$ lies in a plane $[l\ m\ n\ k]$ if and only if

$$Xl + Ym + Zn + Wk = 0.$$

The plane $[l\ m\ n\ k]$ through three non-collinear points is given by the coefficients of X, Y, Z and W in the expansion of the determinant

$$\begin{vmatrix} X & Y & Z & W \\ X_1 & Y_1 & Z_1 & W_1 \\ X_2 & Y_2 & Z_2 & W_2 \\ X_3 & Y_3 & Z_3 & W_3 \end{vmatrix}.$$

In particular, the plane faces of the reference tetrahedron are [1000], [0100], [0010] and [0001].

Similarly, the common point $(X\ Y\ Z\ W)$ of three planes that do not contain a common line is given by the coefficients l, m, n, and k in the expansion of

$$\begin{vmatrix} l & m & n & k \\ l_1 & m_1 & n_1 & k_1 \\ l_2 & m_2 & n_2 & k_2 \\ l_3 & m_3 & n_3 & k_3 \end{vmatrix}.$$

Four points are coplanar if

$$\begin{vmatrix} X_1 & Y_1 & Z_1 & W_1 \\ X_2 & Y_2 & Z_2 & W_2 \\ X_3 & Y_3 & Z_3 & W_3 \\ X_4 & Y_4 & Z_4 & W_4 \end{vmatrix} = 0$$

and four planes are concurrent if

$$\begin{vmatrix} l_1 & m_1 & n_1 & k_1 \\ l_2 & m_2 & n_2 & k_2 \\ l_3 & m_3 & n_3 & k_3 \\ l_4 & m_4 & n_4 & k_4 \end{vmatrix} = 0.$$

2.3 Collineations

Let us write the coordinates $(X\ Y\ Z)$ of a point P in a projective plane as a *column* (i.e., a 3×1 matrix) P and the coordinates $[l\ m\ n]$ of a line L as a row (i.e., a 1×3 matrix) L. The condition for the point P to lie on the line L is then simply $LP = 0$. If T is any non-singular 3×3 matrix, we can apply the transformation $P \to P' = TP$, $L \to L' = LT^{-1}$. Since $L'P' = 0$ the transformation, applied to all points and lines of the projective plane, is a mapping of the plane onto itself, or a mapping of one plane onto another, that *preserves incidences*. It is a *projective transformation*, or *collineation*. Any figure in the plane and its image under a collineation are *projectively equivalent*. Since an overall factor λ is irrelevant for homogeneous coordinates, we may take the matrices T to be unimodular, $|T| = 1$. Then we can say that all the collineations of a real projective plane constitute the special linear group SL(3, R) of all real unimodular matrices.

In general, a collineation T will leave some points unchanged. These are given by the *eigenvectors* of the matrix T, which are the points whose coordinates satisfy

$$TP = \lambda P$$

for some λ. These are the *fixed points* of the collineation. In the general case they will be three non-collinear points. A line through two fixed points is an *invariant* line. If two of the *eigenvalues* λ happen to be equal, then any linear combination of the two corresponding eigenvectors is also an eigenvector, and we get a line of fixed points—a *fixed line*. There are several cases to consider. Collineations may be classified according to their pattern of fixed points and invariant lines. By appropriate choice of the vertices and edges of the reference triangle, relating them to the fixed points and invariant lines, the matrix T takes on simple *canonical forms*:

(a) The general case: the points (100), (010) and (001) are fixed points and the lines [100]. [010] and [001] are invariant lines;

$$T = \begin{pmatrix} \mu & & \\ & \nu & \\ & & \rho \end{pmatrix}, \quad T^{-1} = \begin{pmatrix} \mu^{-1} & & \\ & \nu^{-1} & \\ & & \rho^{-1} \end{pmatrix}.$$

(b) A fixed point (001) and a fixed line [001];

$$T = \begin{pmatrix} \mu & & \\ & \mu & \\ & & \rho \end{pmatrix}, \quad T^{-1} = \begin{pmatrix} \mu^{-1} & & \\ & \mu^{-1} & \\ & & \rho^{-1} \end{pmatrix}.$$

(c) A fixed line [010] and an invariant line [001];

$$T = \begin{pmatrix} \mu & & \\ 1 & \mu & \\ & & \rho \end{pmatrix}, \quad T^{-1} = \begin{pmatrix} \mu^{-1} & -1 & \\ & \mu^{-1} & \\ & & \rho^{-1} \end{pmatrix}.$$

(d) A fixed line [100] and all lines through the point (001) on it invariant;

$$T = \begin{pmatrix} \mu & & \\ 1 & \mu & \\ & & \mu \end{pmatrix}, \quad T^{-1} = \begin{pmatrix} \mu^{-1} & -1 & \\ & \mu^{-1} & \\ & & \mu^{-1} \end{pmatrix}.$$

(e) A fixed point (001) and an invariant line [100] through it;

$$T = \begin{pmatrix} \mu & & \\ 1 & \mu & \\ & 1 & \mu \end{pmatrix}, \quad T^{-1} = \begin{pmatrix} \mu^{-1} & -1 & \\ & \mu^{-1} & -1 \\ & & \rho^{-1} \end{pmatrix}.$$

(f) Finally, we have the trivial case of the collineation that changes nothing: all points and lines are fixed and T is just (a multiple of) the unit matrix.

In Fig. 2.2 fixed points and lines are indicated in black, invariant lines in grey.

Obviously, all this is readily generalizable to collineations in N-dimensional projective space. But it gets complicated!

Fig. 2.2 Classification of plane collineations

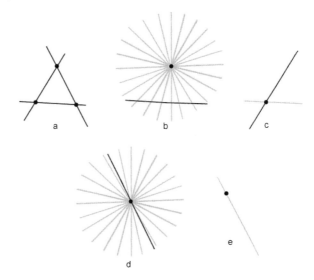

2.4 A Proof of Pappus's Theorem

We are now ready to look at a proof of Pappus's theorem, using the methods of homogeneous coordinates. There are several ways of going about it. In Fig. 2.3 there are two sets of three collinear points ABC and DEF. Let these two lines be respectively the sides [001] and [100] of the reference triangle (i.e. the lines $Z=0$ and $X=0$). The coordinates of the six points then have the form

$$A = (1 \quad \alpha_1 \quad 0), \qquad D = (0 \quad \beta_1 \quad 1)$$
$$B = (1 \quad \alpha_2 \quad 0), \qquad E = (0 \quad \beta_2 \quad 1)$$
$$C = (1 \quad \alpha_3 \quad 0), \qquad F = (0 \quad \beta_3 \quad 1).$$

Fig. 2.3 Pappus's theorem

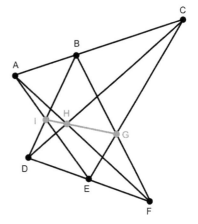

The coordinates of the lines BF and CE are then

$$\begin{vmatrix} 1 & \alpha_2 & 0 \\ 0 & \beta_3 & 1 \end{vmatrix} = [\alpha_2 \;\; -1 \;\; \beta_3] \quad \text{and} \quad \begin{vmatrix} 1 & \alpha_3 & 0 \\ 0 & \beta_2 & 1 \end{vmatrix} = [\alpha_3 \;\; -1 \;\; \beta_2],$$

so that

$$G = \begin{vmatrix} \alpha_2 & -1 & \beta_3 \\ \alpha_3 & -1 & \beta_2 \end{vmatrix} = (\beta_3 - \beta_2, \; \beta_3\alpha_3 - \alpha_2\beta_2, \; \alpha_3 - \beta_2).$$

Similarly,

$$H = (\beta_1 - \beta_3, \; \beta_1\alpha_1 - \alpha_3\beta_3, \; \alpha_1 - \alpha_3) \quad \text{and} \quad I = (\beta_2 - \beta_1, \; \beta_2\alpha_2 - \alpha_1\beta_1, \; \alpha_2 - \alpha_1).$$

Therefore, G, H and I are collinear, because

$$\begin{vmatrix} \beta_3 - \beta_2 & \beta_3\alpha_3 - \alpha_2\beta_2 & \alpha_3 - \alpha_2 \\ \beta_1 - \beta_3 & \beta_1\alpha_1 - \alpha_3\beta_3 & \alpha_1 - \alpha_3 \\ \beta_2 - \beta_1 & \beta_2\alpha_2 - \alpha_1\beta_1 & \alpha_2 - \alpha_1 \end{vmatrix} = 0$$

(the sum of the three rows vanishes).

Notice something peculiar and subtle here. We have tacitly assumed that the numbers used as coordinates *commute*, $\alpha\beta = \beta\alpha$, otherwise the three rows of this determinant would not sum to zero and Pappus's theorem would fail. That is not a problem because we've been taking it for granted that the coordinates are real numbers. The geometry is the 'projective geometry of two dimensions over the reals', P(2, R). But geometries can be constructed *that satisfy the projective geometry axioms*, using coordinates from other number fields, such as complex numbers or finite 'Galois' fields (e.g. numbers modulo p). We then have the projective geometries P(N, C) and P(N, GF(p)). Commutation is still valid for these fields, so Pappus's theorem remains valid. However, for the *quaternions*, for example, commutation fails, so that Pappus's theorem would not be true in P(N, Q).

2.5 Proofs of Desargues' Theorem

Proving Desargues' theorem by means of homogeneous coordinates is very simple. In Fig. 2.4, ABC and DEF are two triangles perspective from a point V. Choose ABC as the reference triangle and V as the unit point. Then D is on VA, so its coordinates are a linear combination of (111) and (100); since homogeneous coordinates can have an arbitrary overall factor we can take them to be of the form (α 1 1). Similarly, E and F can be taken to be (1 β 1) and (1 1 γ). Now BC is [100] and EF is

$$\begin{vmatrix} 1 & 1 & \gamma \\ 1 & \beta & 1 \end{vmatrix} = [1 - \beta\gamma \;\; \gamma - 1 \;\; \beta - 1].$$

2.5 Proofs of Desargues' Theorem

Fig. 2.4 Proof of Desargues' theorem

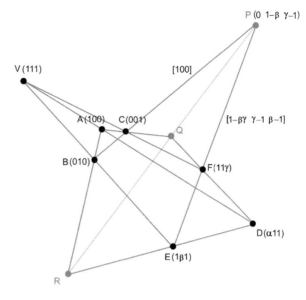

Then P is

$$\begin{vmatrix} 1 & 0 & 0 \\ 1-\beta\gamma & \gamma-1 & \beta-1 \end{vmatrix} = [0 \quad 1-\beta \quad \gamma-1].$$

Similarly, Q is $[\alpha \; -1 \; 0 \; 1-\gamma]$ and R is $[1-\alpha \; \beta-1 \; 0]$. Therefore P, Q and R are collinear because

$$\begin{vmatrix} 0 & 1-\beta & \gamma-1 \\ \alpha-1 & 0 & 1-\gamma \\ 1-\alpha & \beta-1 & 0 \end{vmatrix} = 0.$$

A very elegant proof of the theorem, that does not depend on a particular choice of reference system, is the following. If ABC and DEF are two triangles perspective from V, the homogeneous coordinates of these seven points can be adjusted by multiplying them by arbitrary overall factors so that

$$A + D + V = 0$$
$$B + E + V = 0$$
$$C + F + V = 0.$$

Subtracting these equations in pairs gives

$$A + D = B + E = C + F.$$

Therefore

$$B - C = F - E, \quad C - A = D - F, \quad A - B = E - D.$$

These are the three points P, Q and R in the figure. They are collinear because

$$(B - C) + (C - A) + (A - B) = 0.$$

2.6 Affine Coordinates

We have already observed that an affine plane can be regarded as a specialization of a projective plane in which a line has been singled out, regarded as special, and called the 'line at infinity'. This allows the reintroduction of the affine concept of 'parallel lines'—they are lines that intersect 'at infinity'. Three-dimensional affine space can similarly be treated as a specialization of a projective three-space with a chosen 'plane at infinity', and so on, for higher dimensions.

An affine coordinate system for an affine space derived in this way can be readily obtained from the homogeneous coordinates of the projective space. Take for example the projective plane $P(2, R)$ (the generalization to $P(N, F)$ is obvious). Let (X Y Z) be homogeneous coordinates for the points of the projective plane. Choose the coordinate system so that the 'line at infinity' is $Z = 0$, that is, [0 0 1]. Then all the finite points in the affine plane have $Z \neq 0$, and the homogeneous coordinates for the affine plane can be taken to be

$$(X, Y) = (X/Z, Y/Z)$$

the line [l m n] becomes the line with equation

$$lX + mY + n = 0.$$

It is often convenient and fruitful to employ the converse of this procedure, making use of homogeneous coordinates to deal with problems in affine geometry. (An example will be demonstrated in Sect. 3.4.)

2.7 Subspaces of a Vector Space

In the above construction of homogeneous coordinates, described in Sect. 2.1, *any* plane p that does not pass through E can be used—all the planes that do not contain E are projectively equivalent. They are related by collineations. It follows that we can remove the plane p from Fig. 2.1 and we still have a plane projective geometry if we simply *call* the lines through E 'points' and the planes through E 'lines'. The axioms of plane projective geometry are valid for these 'points' and 'lines'. In fact, we are dealing with the three-dimensional *vector space* of position vectors, with *E* as origin. The 'points' are its one-dimensional subspaces and the 'lines' are its two-dimensional subspaces. We can therefore say that a one-dimensional *vector space* is a zero-dimensional *projective* space (a 'point') and that a two-dimensional vector space is a one-dimensional projective space (a 'line'). This can be immediately generalized to a definition of higher-dimensional projective geometries:

2.7 Subspaces of a Vector Space

The set of all vector subspaces of an $(N+1)$-dimensional vector space is an N-dimensional projective space.

The *union* of two vector subspaces is the smallest subspace that contains them both and their *intersection* is the largest subspace that is contained in both. The union and intersection of two vector subspaces then correspond, respectively, to the join and intersection of the corresponding subspaces of a projective space. It is then a simple matter to deduce that *the axioms for N-dimensional projective space that we presented in Sect. 1.10 are all simple consequences of the properties of a vector space.*

We need now to review briefly some of the basic ideas of vector space theory.

A *vector space* V over the field R of real numbers is the set of all vectors of the form

$$x^1\mathbf{e}_1 + x^2\mathbf{e}_2 + x^3\mathbf{e}_3 + \cdots + x^{N+1}\mathbf{e}_{N+1} = x^i\mathbf{e}_i$$

where the x^i are real numbers and the $\mathbf{e}_i$ are $(N+1)$ vectors, which are *linearly independent* but otherwise arbitrary. (By linear independence we mean that the above expression can be zero only if all the x^i are zero.) The set of vectors $\mathbf{e}_i$ is a *basis* for V. The indices i ($i = 1, 2, \ldots, N+1$) in the shorthand expression $x^i\mathbf{e}_i$ have been written as superscripts on the x^i so that we can use the *summation convention*: if an index appears twice in an expression, once as a subscript and once as a superscript, a summation is implied. (This convention, introduced by Einstein, allows us to omit summation signs $\sum$ and simplifies the look of many expressions in mathematics and theoretical physics.)

A one-dimensional subspace of V is the set of all vectors of the form $\lambda \mathbf{v}$, for *any* real number λ and some fixed vector $\mathbf{v}$. This is a *point* in the N-dimensional projective space $P(N, R)$. The components x^i of $\mathbf{v}$ are its homogeneous coordinates, as also are λx^i for any non-zero real number λ. Similarly, a line in $P(N, R)$ is given by all linear combinations $\lambda \mathbf{v} + \mu \mathbf{w}$ of two linearly independent fixed vectors ('there is a unique line through any two distinct points'...), a plane is given by all linear combinations of three linearly independent fixed vectors, and so on.

In the projective plane $P(2, R)$, we denoted the homogeneous coordinates of points by $(X\ Y\ Z)$ and introduced homogeneous coordinates $[l\ m\ n]$ of a line in such a way that the condition for the point to lie on the line is $lX + mY + nZ = 0$. This readily generalizes: in $P(N, R)$, we introduce homogeneous coordinates $[\omega_1, \omega_2, \ldots, \omega_{N+1}]$ for a hyperplane (($N-1$)-dimensional subspace), so that the condition for the point $(x^1, x^2, \ldots, x^{N+1})$ to lie in the hyperplane is $\omega_i x^i = 0$. In terms of the vector space V, this suggests that the hyperplanes can be thought of as *dual* vectors. We can associate, with any vector space V, a *dual* vector space V^*. A vector $\mathbf{v}$ in V and a vector $\boldsymbol{\omega}$ in V^* can be combined to give a scalar (in this case, a real number) $\boldsymbol{\omega}\mathbf{v}$, and this product preserves linearity, in the sense that $(\lambda\boldsymbol{\omega})\mathbf{v} = \boldsymbol{\omega}(\lambda\mathbf{v}) = \lambda(\boldsymbol{\omega}\mathbf{v})$, $(\boldsymbol{\omega}_1 + \boldsymbol{\omega}_2)\mathbf{v} = \boldsymbol{\omega}_1\mathbf{v} + \boldsymbol{\omega}_2\mathbf{v}$ and $\boldsymbol{\omega}(\mathbf{v}_1 + \mathbf{v}_2) = \mathbf{v}_1 + \mathbf{v}_2$, where λ is any real number, $\boldsymbol{\omega}_1$ and $\boldsymbol{\omega}_2$ are any two vectors in V^* and $\mathbf{v}_1 + \mathbf{v}_2$ are any two vectors in V. A basis $\mathbf{e}^i$ for V^* can be chosen so that

$$\mathbf{e}^i\mathbf{e}_j = \delta^i_j$$

(δ^i_j is defined to be 1 if $i = j$, otherwise 0). Then, if $\boldsymbol{\omega} = \omega_i \mathbf{e}^i$ and $\mathbf{v} = x^i \mathbf{e}_i$

$$\boldsymbol{\omega}\mathbf{v} = \omega_i x^i.$$

It follows that, just as the points in P(N, R) can be identified with one-dimensional subspaces of V, the hyperplanes of P(N, R) can be identified with one-dimensional subspaces of V*, and that the condition for a point $\mathbf{v}$ to lie in a hyperplane $\boldsymbol{\omega}$ is $\boldsymbol{\omega}\mathbf{v} = 0$.

2.8 Plücker Coordinates

In a three-dimensional projective space (P(3, R), say), points are assigned four homogeneous coordinates $(x^1\ x^2\ x^3\ x^4)$, planes are assigned four homogeneous coordinates $[\pi_1\ \pi_2\ \pi_3\ \pi_4]$, and the condition for a point x^i to lie in a plane π_i is $x^i \pi_i = 0$ (where a summation is implied over the repeated index i). In this notation, coordinates of points are indexed with a superscript and coordinates of planes are indexed with a subscript. The significance of this convention is that it indicates the nature of the change of coordinates under the action of a collineation (given by a 4×4 matrix T):

$$x^i \rightarrow T^i_j x^j, \qquad \pi_i \rightarrow (T^{-1})^j_i \pi_j.$$

The *lines* of a three-dimensional space can be assigned a set of *six* homogeneous coordinates. If x^i and y^i are any two distinct points on a line P, the *Plücker coordinates* of P are

$$P^{ij} = -P^{ji} = x^i y^j - x^j y^i.$$

Observe that it does not matter which two points on P are chosen. With a different choice we get, apart from an irrelevant overall factor, the same set of Plücker coordinates. Defining $x'^i = \alpha x^i + \beta y^i$, $y'^i = \gamma x^i + \delta y^i$, then $P'^{ij} = (\alpha\delta - \beta\gamma)P^{ij}$.

It is not difficult to see that the condition for a line P to lie in a plane π is

$$P^{ij}\pi_j = 0$$

and that if this condition is not satisfied, $P^{ij}\pi_j$ gives the coordinates of the point of intersection of the line P with the plane π.

The P^{ij} defined in this way satisfy

$$P^{23}P^{14} + P^{31}P^{24} + P^{12}P^{34} = 0.$$

Moreover, this is a necessary and sufficient condition for a skewsymmetric P^{ij} to be a set of Plücker coordinates of a line—that is, for P^{ij} to have the form $x^i y^j - x^j y^i$. To see this, take two planes π and μ such that $x^i = P^{ij}\pi_j$ and $y^i = P^{ij}\mu_j$ are nonzero. We assume, of course, that the P^{ij} are not all zero, so we lose no generality by

2.9 Grassmann Coordinates

taking $P^{34} = 1$ and choosing the reference system so that π and μ are the reference planes [0010] and [0001]. Then

$$(x^1 \quad x^2 \quad x^3 \quad x^4) = (P^{13} \quad P^{23} \quad 0 \quad -1) \quad \text{and}$$

$$(y^1 \quad y^2 \quad y^3 \quad y^4) = (P^{14} \quad P^{24} \quad 1 \quad 0).$$

Hence $Q^{ij} = x^i y^j - x^j y^i$ is given by the skewsymmetric matrix

$$\begin{pmatrix} 0 & P^{13}P^{24} - P^{23}P^{14} & P^{13} & P^{14} \\ P^{23}P^{14} - P^{13}P^{24} & 0 & P^{23} & P^{24} \\ P^{31} & P^{32} & 0 & P^{34} \\ P^{41} & P^{42} & P^{43} & 0 \end{pmatrix}$$

applying $P^{ij} = -P^{ji}$ and the condition $P^{23}P^{14} + P^{31}P^{24} + P^{12}P^{34} = 0$ then gives $Q^{ij} = P^{ij}$.

The same line P can be given *dual* Plücker coordinates, by taking any two distinct planes π_i and μ_i whose intersection is the line P:

$$P_{ij} = \pi_i \mu_j - \pi_j \mu_i.$$

These two kinds of coordinate for a line are related through

$$P_{23} = P^{14}, \qquad P_{31} = P^{24}, \qquad P_{12} = P^{34},$$

$$P_{14} = P^{23}, \qquad P_{24} = P^{31}, \qquad P_{34} = P^{12}.$$

This is more briefly expressed if we introduce the *alternating symbol*

$\varepsilon^{ijkl} = \varepsilon_{ijkl} = \pm 1$ according as $ijkl$ is an even or an odd permutation of 1234, otherwise zero.

Then

$$P_{ij} = \left(\frac{1}{2}\right)\varepsilon_{ijkl}P^{kl}, \qquad P^{ij} = \left(\frac{1}{2}\right)\varepsilon^{ijkl}P_{kl}.$$

The condition given above for a skewsymmetric P^{ij} to be Plücker coordinates of a line is then $P^{ij}P_{ij} = 0$. The condition for a point x to lie on a line P is $P_{ij}x^j = 0$, the condition for a line P to be contained in a plane π is $P^{ij}\pi_j = 0$. If a plane λ_i does not contain the line P ($P^{ij}\lambda_j \neq 0$), then $x^i = P^{ij}\lambda_j$ is the point of intersection of the line and the plane. The condition $P^{23}P^{14} + P^{31}P^{24} + P^{12}P^{34} = 0$ for a given skewsymmetric P^{ij} to be a set of Plucker coordinates of a line is $P^{ij}P_{ij} = 0$. The condition for two lines P and Q to intersect is $P^{ij}Q_{ij} = 0$.

2.9 Grassmann Coordinates

Two questions arise. When deriving the projective geometry P(N, R) from an $(N + 1)$-dimensional vector space V, we have introduced two quite different definitions of a hyperplane: as an N-dimensional subspace of V, or as a one-dimensional subspace of V*. How are these apparently different definitions to be reconciled? And how can one assign homogeneous coordinates for lines, planes, etc., in a projective space of more than two dimensions? The generalization of Plücker's coordinates for lines in 3-space provides the answer to these questions.

The n-dimensional subspace determined by $n+1$ general points of an N-dimensional projective space has a set of $\binom{N+1}{n}$ homogeneous coordinates.

To investigate this general case we introduce the *alternating symbols*

$$\varepsilon_{i_1 i_2 i_3 \ldots i_{N+1}} = \varepsilon^{i_1 i_2 i_3 \ldots i_{N+1}} = \pm 1$$ according as $i_1 i_2 i_3 \ldots i_{N+1}$ is an even or an odd permutation of $123, \ldots, N+1$, otherwise zero.

Now, if x^i and y^i ($i = 1, 2, \ldots, N+1$) are the coordinates of two distinct points, the $\binom{N+1}{2}$ numbers

$$\omega^{ij} = x^{[i} y^{j]} = \frac{1}{2}\left(x^i y^j - x^j y^i\right)$$

are not all zero if the points are distinct, and serve to specify the unique line through the two points. We can adopt the ω^{ij} as the homogeneous coordinates of the line through the two points. Similarly, given three non-collinear points with coordinate x^i, y^i and z^i, we can take as homogeneous coordinates of the plane through these three points the $\binom{N+1}{3}$ numbers

$$\omega^{ijk} = x^{[i} y^j z^{k]} = (1/6)\left(x^i y^j z^k + x^j y^k z^i + x^k y^i z^j - x^i y^k z^j - x^k y^j z^i - x^j y^i z^k\right).$$

And so on. The alternating symbols allow us to construct alternative sets of homogeneous coordinates, the *dual* coordinates. Dual coordinates of points, lines, planes, etc., are (using the summation convention) given by

$$x_{i_1 i_2 i_3 \ldots i_N} = \varepsilon_{i_1 i_2 i_3 \ldots i_N j} x^j$$

$$\omega_{i_1 i_2 i_3 \ldots i_{N-1}} = (1/2)\varepsilon_{i_1 i_2 i_3 \ldots i_{N-1} jk} \omega^{jk}$$

$$\omega_{i_1 i_2 i_3 \ldots i_{N-2}} = (1/6)\varepsilon_{i_1 i_2 i_3 \ldots i_{N-2} jkl} \omega^{jkl}$$

and so on. The properties of the alternating symbol imply

$$x^j = (1/N!) x_{i_1 i_2 i_3 \ldots i_N} \varepsilon^{i_1 i_2 i_3 \ldots i_N j}$$

$$\omega^{jk} = (1/(N-1)!) \omega_{i_1 i_2 i_3 \ldots i_{N-1}} \varepsilon^{i_1 i_2 i_3 \ldots i_{N-1} jk}$$

$$\omega^{jk} = (1/(N-2)!) \omega_{i_1 i_2 i_3 \ldots i_{N-2}} \varepsilon^{i_1 i_2 i_3 \ldots i_{N-2} jkl}$$

et cetera. These expressions are simpler than they look. Taking $N = 4$ as an example, they are just $x_{1234} = x^5$, $\omega_{123} = \omega^{45}$ and $\omega_{12} = \omega^{345}$ and the expressions obtained from these by cyclically permuting 12345. For $N = 3$ we have the Plücker coordinates of a line.

2.9 Grassmann Coordinates

(Readers who are familiar with the algebra of 'exterior forms' will recognize all this. A 'one-form' is a dual vector $\omega_i \mathbf{e}^i$. A basis for 2-forms $\omega_{ij}\mathbf{e}^{ij}$ is a set of symbolic quantities $\mathbf{e}^{ij}$ $(= -\mathbf{e}^{ji})$, usually written as $\mathbf{e}^i \wedge \mathbf{e}^j$. For 3-forms $\omega_{ijk}\mathbf{e}^{ijk}$ we have $\mathbf{e}^{ijk}$ $(= \mathbf{e}^i \wedge \mathbf{e}^j \wedge \mathbf{e}^k)$, completely skewsymmetric in the index set ijk. *Et cetera.* Dualizing with the alternating symbol corresponds to the Hodge $*$ operation.)

We shall have no further use for Grassman's algebraic ideas in what follows— they have been introduced here simply to hint at how the principle of duality in N-dimensional projective geometry can be expressed algebraically in terms of homogeneous coordinates. In particular, we see now how a hyperplane in P(N, R) can be regarded without contradiction either as an N-dimensional subspace of a vector space V or as a one-dimensional subspace of V*.

Chapter 3
Linear Figures

Abstract An important projective invariant, the cross-ratio of four collinear points, is defined. Some special configurations in two, three and four dimensions are introduced, including extensions of Desargues' theorem, Sylvester's 'duads and synthemes', desmic systems of tetrahedra, and Baker's remarkable configuration derived from six points in a complex four-dimensional space.

3.1 The Projective Line

Even one-dimensional projective spaces are not trivial. By analogy from what we have said above about homogeneous coordinates, it is clear that a point in a one-dimensional projective space can be assigned a pair of homogeneous coordinates $(X\ Y)$. Since all the coordinate pairs $\lambda(X\ Y) = (\lambda X\ \lambda Y)$, for any non-zero λ, refer to the same point, the ratio $\theta = X/Y$ is sufficient to identify a point. This is a *homographic parameter* for the points on the line. Projective transformations on a line are called *homographies*. We must include the point $\theta = \infty$, corresponding to $Y = 0$, as the parameter of a valid point—otherwise, we would have an affine line rather than a projective line. (Alternatively, by analogy with the assignment of homogeneous coordinates for the lines in a projective plane, we could label points by 'dual' coordinates $[l\ m] = [Y\ -X]$. The condition for two points to coincide is then $lX + mY = 0$.) Just as the group of projective transformations on the real projective plane is SL(3, R), the group of projective transformation on the real projective line is the group SL(2, R) of all real unimodular 2×2 matrices:

$$T = \begin{pmatrix} a & b \\ c & d \end{pmatrix},$$

satisfying $|T| = 1$.

In terms of the homographic parameter, these transformations are

$$\theta \to \theta'' = \frac{a\theta + b}{c\theta + d}, \quad ad - bc = 1$$

(including $\infty \to a/c$ and $-d/c \to \infty$). Note that any three points A, B and C on a projective line can be chosen to be the *reference points* (10), (01) and (11); that is,

Fig. 3.1 Reference points on a projective line

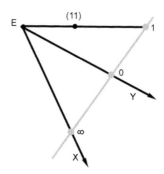

they can be assigned the homographic parameters ∞, 0 and 1. This is just a lower-dimensional analogue of what we showed in Fig. 2.1: it is a matter of establishing homogeneous coordinates on the line by choosing the axes and the point (11) in an affine plane, as in Fig. 3.1.

A non-trivial homography leaves two points of the line fixed.

(The 'trivial' homography is $\theta' = \theta$, which leaves every point on the line fixed.) As already noted, a homography maps ∞ to a/c, and so, if $c = 0$, $\theta = \infty$ is a fixed point and the other fixed point is $\theta = b/(d-a)$. If $c \neq 0$ the fixed points are given by the roots of

$$c\theta^2 + (d-a)\theta - b = 0.$$

They may be real and distinct, real and coincident, or conjugate complex, according to whether the discriminant

$$(d-a)^2 - 4bc = (d-a)^2 - 4$$

is positive, zero or negative.

The fixed points of a homography $\theta \to \theta'' = \frac{a\theta+b}{c\theta+d}$ *are real and distinct, real and coincident, or conjugate complex, according to whether the eigenvalues of* $T = \begin{pmatrix} a & b \\ c & d \end{pmatrix}$ *are real and distinct, real and coincident, or conjugate complex.*

Proof The eigenvalues of T are the roots of

$$\begin{vmatrix} a-\lambda & b \\ c & d-\lambda \end{vmatrix} = \lambda^2 - (a+d)\lambda + 1 = 0.$$

The discriminant is the same, $(a+d)^2 - 4$. □

A homography is uniquely determined if three distinct points and their three images are given. To show this algebraically we can choose the three points to be

∞, 0 and 1 and their respective images to be θ_1, θ_2 and θ_3. Then $a/c = \theta_1$, $b/d = \theta_2$ and $(a+b)/(c+d) = \theta_3$. A little tedious algebra then gives

$$\begin{pmatrix} a & b \\ c & d \end{pmatrix} = \frac{1}{(\theta_1 - \theta_2)(\theta_2 - \theta_3)(\theta_3 - \theta_1)} \begin{pmatrix} \theta_1(\theta_3 - \theta_2) & \theta_2(\theta_1 - \theta_3) \\ \theta_3 - \theta_2 & \theta_1 - \theta_3 \end{pmatrix}.$$

3.2 Cross-Ratio

Now let θ_1, θ_2, θ_3 and θ_4 be the parameters for four points A, B, C and D on the line. It is easily verified that

$$(AB, CD) = (\theta_1 - \theta_3)(\theta_2 - \theta_4)/(\theta_1 - \theta_4)(\theta_2 - \theta_3)$$

does not change when a homography is applied. It is a *projective invariant*. It is called the *cross-ratio* of the two point-pairs AB and CD.

The dual, in a projective plane, of four collinear points is, of course, four concurrent lines. Hence a homographic parameter can label all the lines through a point, and the cross-ratio of four concurrent lines is defined in an obvious way. (In three dimensions, we can similarly define the cross-ratio for four planes through a line.) The cross-ratio of four concurrent lines is then equal to the cross-ratio of their four points of intersection with any other line that does not contain their point of intersection. The proof of this is fairly obvious so I shall leave it as a simple exercise.

Now let A, B, C and D be the $(N+1)$-tuples of homogeneous coordinates of four *collinear* points in a projective N-space. Then C and D are linear combinations of A and B, and by adjusting the arbitrary overall scalar factors, we can write $C = A + B$ and $D = \chi A + B$. Then

$$\chi = (AB, CD),$$

because if we choose A, B and C to be the reference points and unit point on the line, with homographic parameters ∞, 0 and 1, the homographic parameter of D is then χ. Cross-ratios are invariant under projective transformations of the N-space.

On any line, two points C and D are *harmonic conjugates* of each other with respect to the point pair AB (and A and B are harmonic conjugates with respect to the point pair CD) if and only if $(AB, CD) = -1$. This can be proved by taking the triangle ABH in Fig. 1.10 to be the reference triangle: $A = (100)$, $B = (010)$ and $H = (001)$. The lines BH, HA and AB are then [100], [010] and [001]. Let the unit point (111) be F. Then FA is [0 1 −1], FB is [−1 0 1] and FH is [1 −1 0]. So then the intersection C of FH and AB is (110), the intersection G of BH and FA is (011) and the intersection I of AH and FB is (101). Then GI is [1 1 −1]. Finally, we find that D, the intersection of AB and GI, is (−1 1 0). The ratios X/Y for the four points A, B, C and D on the line [001] are then $\theta_1 = \infty$, $\theta_2 = 0$, $\theta_3 = 1$ and $\theta_4 = -1$. Therefore $(AB, CD) = (CD, AB) = -1$.

Fig. 3.2 An involutory hexad

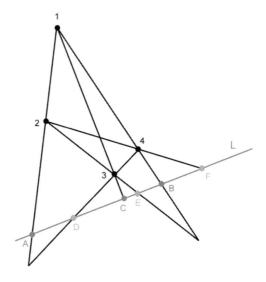

3.3 Involutions

If the matrix T of a projective transformation satisfies $T^2 = \pm I$ (where I is the unit matrix), then the points (apart from the fixed points) of the projective space are associated in pairs: if T takes the point P to P′, it takes P′ to P. A projective transformation with this property is called an *involution*. (In Euclidean geometry, *reflections and half-turns* are examples of involutions.) The condition for a homography on a projective line to be an involution is $a = -d$.

Six points ABCDEF on a projective line are said to constitute an *involutory hexad* (ABC, DEF) if there is a *unique involution* that maps A to D, B to E and C to F.

A remarkable theorem states that *A general line* L *intersects the six lines of a complete quadrangle in the six points of an involutory hexad.*

In Fig. 3.2 the four vertices of a complete quadrangle in a projective plane are labeled by the symbols **1234**. Its six lines cut a line **L** in the points **ABCDEF**. Choose **AB1** as the reference triangle and **3** as the unit point. That is, **A** = (100), **B** = (010), **1** = (001) and **3** = (111). Thus **L** = **AB** is [001]. The line **13** is [1 −1 0] which intersects [001] at **C** = (110). Homographic parameters for **A**, **B** and **C** are therefore, respectively, ∞, 0 and 1. The point **2** lies on the line **1A**, which is [010] and so **2** has the form $(\mu\ 0\ 1)$. Similarly, **4** lies on the line **1B**, which is [100] and so **4** has the form $(0\ \nu\ 1)$. Then **43** is $[\nu - 1\ 1\ -\nu]$ which intersects **L** at **D** = $(1\ 1 - \nu\ 0)$. Also, **23** is $[-1\ 1 - \mu\ \mu]$, which intersects **L** at **E** = $(1 - \mu\ 1\ 0)$. Finally, **24** is $[\nu\ \mu\ -\nu\ \mu]$ which intersects **L** at **F** = $(\mu\ -\nu\ 0)$. We now have homographic parameters for **D**, **E** and **F**. They are

$$\theta_1 = 1/(1-\nu), \qquad \theta_2 = 1 - \mu, \qquad \theta_3 = -\mu/\nu$$

3.4 Cross-Ratio in Affine Geometry

As we saw in the previous section, the homography that maps ∞, 0 and 1 to θ_1, θ_2 and θ_3, respectively, has

$$\alpha/\delta = \theta_1(\theta_3 - \theta_2)/(\theta_1 - \theta_3).$$

In this case, this turns out to be -1. Hence $\alpha = -\delta$. The homography is an involution. (**ABC**, **DEF**) is an involutory hexad.

3.4 Cross-Ratio in Affine Geometry

In affine geometry, the 'length' of a line segment AB has no meaning. That is a Euclidean concept. It is, however, possible to define a length *ratio* AB/CD of two line segments AB and CD on the same line, or on two parallel lines. Let A, B and C be three collinear points in a projective space and choose a hyperplane at infinity for specializing to an affine space. Let Ω be the intersection of the line ABC with the hyperplane at infinity. Define

$$AC/BC = -AC/CB = CA/CB = -CA/BC = (AB, C\Omega).$$

Note, incidentally, that a consequence of this definition of AC/BC for three points A, B and C on an affine line is that:

> *The mid-point* C *of a line segment* AB *is the harmonic conjugate of the point at infinity on the line* AB.

(Without loss of generality a homographic parameter on the line AB can be chosen so that A, B, C and Ω are assigned the parameters 0, 1, θ and ∞, respectively. We find that AB/BC $= -1$ when $\theta = 1/2$.)

Since cross-ratio is a projective invariant, the quantity AC/AB defined in this way is an affine invariant—invariant under collineations that leave the line at infinity invariant. More generally, for four collinear points A, B, C and D,

$$AC/BD = (AC/BC)(BC/BD).$$

When AB and BD are on two parallel lines rather than on the same line, we can simply transfer BD to B'D' on the line AB, such that BB' $\parallel$ DD' (that is, *by definition*, 'opposite sides of a parallelogram are equal', BD/B'D' $= 1$).

As an example of the use of the concept of cross-ratio in affine geometry, consider Fig. 3.3. Choose the triangle ABC as the reference triangle, and GHI as the unit line [111]. Let the line DEF be [l m n]. Then D, E and F are respectively (0 n $-m$), ($-n$ 0 l) and (m $-l$ 0), and G, H and I are (0 1 -1), (-1 0 1) and (1 -1 0).

Now, adjusting arbitrary overall factors, so that $G = B + C$ we have $D = (n/m)B + C$, and hence (BC, DG) $= n/m$. Similarly, (CA, EH) $= l/n$ and (AB, FI) $= m/l$. Hence

$$(BC, DG)(CA, EH)(AB, FI) = 1.$$

Fig. 3.3 Projective generalization of Menelaus' theorem

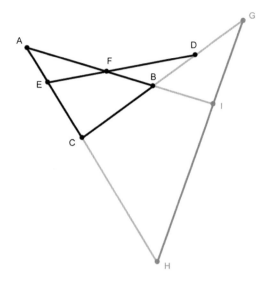

Now specialize to an affine plane by choosing GHI as the 'line at infinity'. We get the theorem of Menelaus as it was known to Euclid:

If a line cuts the sides of a triangle ABC *at points* D, E *and* F, *then* (BC/DB)(CA/EC)(AB/FA) = −1.

Civa's theorem is closely related, and can be proved in a similar way: *if* ABC *is a triangle and* D, E *and* F *are points on the sides* BC, CA *and* AB, *respectively, then* AD, BE *and* CF *are concurrent if and only if* AF/BF · BD/CD · CE/AE = −1.

3.5 The Complex Projective Line

We have been taking it for granted that homogeneous coordinates are rational numbers or real numbers. There was no need for complex numbers because all the expressions concerning incidence properties in projective spaces, that we have been considering so far, have been linear expressions. Complex numbers have an important part to play in projective geometry once we start to consider algebraic curves, surfaces and hypersurfaces other than lines, planes and hyperplanes. The complex projective line P(1, C) or PL(C) provides a simple introduction.

A homogeneous quadratic equation

$$aX^2 + 2bXY + cY^2 = 0$$

is satisfied by two points with homographic parameters θ_1 and θ_2 given by the roots of

$$a\theta^2 + 2b\theta + c = 0$$

3.5 The Complex Projective Line

Of course, even when a, b and c are real θ_1 and θ_2 may be real and distinct, real and equal, or a conjugate complex pair.

The matrix Q of the quadratic form is the symmetric matrix

$$\begin{pmatrix} a & b \\ b & c \end{pmatrix}.$$

The discriminant of the quadratic equation is $b^2 - ac = -|Q|$. The two points are real and distinct, real and coincident, or conjugate complex, according to whether $-|Q|$ is positive, zero or negative.

An interesting variant of these simple ideas is to consider, instead of a symmetric matrix, a *Hermitian* matrix (i.e., a matrix whose transpose is its complex conjugate)

$$Q = \begin{pmatrix} \alpha & \bar{\beta} \\ \beta & \gamma \end{pmatrix}$$

where α and γ are real but β is complex and $\bar{\beta}$ denotes its conjugate. We get a very interesting geometrical interpretation of the equation

$$\alpha \theta \bar{\theta} + \bar{\beta} \theta + \beta \bar{\theta} + \gamma = 0$$

if we interpret complex numbers as points in a two-dimensional space (the usual Argand plane). We have $\theta = \xi + i\eta$, and the plane is augmented by a single point corresponding to $\theta = \infty$. The equation of a circle, centre $(\xi_0 \eta_0)$ and radius R is

$$|\theta - \theta_0|^2 = (\xi - \xi_0)^2 + (\eta - \eta_0)^2 = R^2.$$

Now notice that $\alpha \theta \bar{\theta} + \bar{\beta} \theta + \beta \bar{\theta} + \gamma = 0$ is precisely of this form, with

$$\theta_0 = -\beta/\alpha \qquad R^2 = -|Q|/\alpha^2$$

and if $\alpha = 0$ we get a *line* in the Argand plane. We may regard a line as a 'circle' of infinite radius. Then:

> There is a one-to-one correspondence between the points on the complex projective line that satisfy $\bar{P}^T Q P = 0$ (Q Hermitian, $Q = Q^\dagger = \bar{Q}^T$, and non-singular) *and the circles and lines in the Argand plane.*

As we might expect, a homography on PL(C),

$$\theta \to \frac{a\theta + b}{c\theta + d}$$

(where a, b, c and d may be complex) gives a circle-preserving transformation in the Argand plane. We can combine these homographies with complex conjugation

$$\theta \to \bar{\theta}$$

which just corresponds to a reflection in the Argand plane, so the *anti-homographies*

$$\theta \to \frac{a\bar{\theta}+b}{c\bar{\theta}+d}$$

also lead to circle-preserving transformations.

$$\theta \to 1/\bar{\theta},$$

for example, corresponds to inversion in the unit circle centred at the origin.

3.6 Equianharmonic Points

The cross-ratio of four collinear points $(AB, CD) = (CD, AB) = \chi$ depends on the order of the points. If $(AB, CD) = \chi$, then there are in general six cross-ratios:

$(AB, CD) = \chi$ $\qquad (AB, DC) = 1/\chi$
$(AC, BD) = 1 - \chi$ $\qquad (AC, DB) = 1/(1-\chi)$
$(AD, BC) = 1 - 1/\chi$ $\qquad (AD, CB) = \chi/(1-\chi)$.

If $(AB, CD) = (AC, DB) = (AD, BC)$, then there are only two cross-ratios instead of six:

$$\chi^2 - \chi + 1 = 0$$

and therefore $\chi = -\omega$ or $-\omega^2$, where

$$\omega = (1 + i\sqrt{3})/2$$

is the cube root of unity, which satisfies

$$\omega^3 = 1, \qquad \omega^2 = 1/\omega = \bar{\omega}, \qquad 1 + \omega + \omega^2 = 0.$$

($\bar{\omega}$ denotes the complex conjugate of ω.) The cross-ratio in this case is unchanged by cyclic permutation of any three of the four points. It follows readily that any three *real* collinear points uniquely determine a pair of complex conjugate points on the same line, the *equianharmonic points*. The homographic parameters for the five points can be taken to be $\infty, 0, 1, -\omega$ and $-\bar{\omega}$.

3.7 Four Points in a Plane

Four points in a plane, no three of which are collinear, determine a *complete quadrangle*

3.7 Four Points in a Plane

$$\begin{array}{|cc|} \hline 4 & 3 \\ 2 & 6 \\ \hline \end{array}$$

—a $(4_3 6_2)$ of points and planes. This is what we get if we *project* a tetrahedron (α_3) from a general point in 3-space on to a general plane; if we make a planar *section* of a configuration in 3-space—its intersection with a general plane—the images of its lines are points and the images of its planes are lines. The *section* of a tetrahedron by a general plane is a $(6_2, 4_3)$ of points and lines—a *complete quadrilateral*. This much is trivial, but as we move up to its analogues in higher dimensions we get a sequence of quite amazing configurations.

The homogeneous coordinates of four points **ABCD** in a plane can be chosen to satisfy

$$\mathbf{A} + \mathbf{B} + \mathbf{C} + \mathbf{D} = 0.$$

(The four coordinate triples are necessarily linearly dependent, $\lambda \mathbf{A} + \mu \mathbf{B} + \nu \mathbf{C} + \rho \mathbf{D} = 0$, and since homogeneous coordinates are specified only up to an arbitrary numerical factor, we can choose $\lambda = \mu = \nu = \rho = 1$.) The three *diagonal points* of the complete quadrangle are the intersections of the three pairs of opposite sides. They are given by

$$\mathbf{B} + \mathbf{C} \quad \mathbf{C} + \mathbf{A} \quad \mathbf{A} + \mathbf{C}$$

or, equivalently,

$$\mathbf{A} + \mathbf{D} \quad \mathbf{B} + \mathbf{D} \quad \mathbf{C} + \mathbf{D}.$$

Now, the harmonic conjugate of $\mathbf{A} + \mathbf{B}$ with respect to the point pair **AB** is $\mathbf{A} - \mathbf{B}$, and so on. The six *harmonic points*

$$\begin{array}{ccc} \mathbf{A} - \mathbf{B} & \mathbf{A} - \mathbf{C} & \mathbf{A} - \mathbf{D} \\ & \mathbf{B} - \mathbf{C} & \mathbf{B} - \mathbf{D} \\ & & \mathbf{C} - \mathbf{D} \end{array}$$

are collinear in threes on four lines (because $(\mathbf{A} - \mathbf{B}) + (\mathbf{B} - \mathbf{C}) + (\mathbf{C} - \mathbf{A}) = 0$, etc.), forming a *complete quadrilateral* $(6_3 4_2)$.

All this is shown in Fig. 3.4 (recall Fig. 1.11, which showed the same situation in a different context). In this figure we have done some renaming of the points, for convenience and to set the scene for analogous situations in higher dimensions. The three *diagonal points* $\mathbf{B} + \mathbf{C}, \mathbf{C} + \mathbf{A}$ and $\mathbf{A} + \mathbf{B}$ are denoted by **23 31** and **12** and we have renamed the six *harmonic points*

$$\begin{array}{ccc} 12 & 13 & 14 \\ & 23 & 24 \\ & & 34 \end{array}$$

Observe all the *harmonic ranges*—sets of four collinear points in harmonic relationship. There are nine of them.

Fig. 3.4 Harmonic relationships arising from four points in a plane

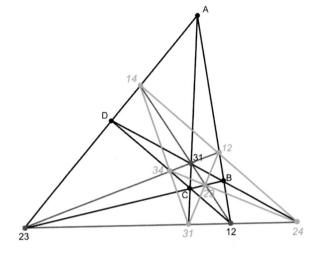

It is instructive to follow the above construction algebraically, in terms of homogeneous coordinates—just as a simple illustration of the manipulation of homogeneous coordinates for points and lines in a projective plane. We can take **ABC** to be the reference triangle, so that $\mathbf{A} = (100)$, $\mathbf{B} = (010)$ and $\mathbf{C} = (001)$. **D** is then the unit point $(-1\ -1\ -1)$. The edges of the reference triangle, **BC**, **CA** and **AB** are [100], [010] and [001].

The line **AD** is then

$$\begin{vmatrix} 1 & 0 & 0 \\ -1 & -1 & -1 \end{vmatrix} = [0\ \ 1\ \ -1].$$

Similarly, $\mathbf{BD} = [-1\ 0\ 1]$ and $\mathbf{CD} = [1\ -1\ 0]$. AD intersects BC at **23**, which is

$$\begin{vmatrix} 0 & 1 & -1 \\ 1 & 0 & 0 \end{vmatrix} = (0\ \ 1\ \ 1).$$

Similarly, $\mathbf{31} = (1\ 0\ 1)$ and $\mathbf{12} = (0\ 1\ 1)$. The edge **23 31** is

$$\begin{vmatrix} 1 & 0 & 1 \\ 0 & 1 & 1 \end{vmatrix} = [-1\ \ 1\ \ 1].$$

It intersects **AB** at

$$\begin{vmatrix} -1 & 1 & 1 \\ 1 & 0 & 0 \end{vmatrix} = (0\ \ 1\ \ -1).$$

This is the harmonic point $\mathbf{B} - \mathbf{C} = \mathbf{23}$. The collinearity of *23 31 12* is shown by the vanishing of the determinant

$$\begin{vmatrix} 0 & 1 & -1 \\ -1 & 0 & 1 \\ 1 & -1 & 0 \end{vmatrix}$$

3.8 Configurations in More than Two Dimensions

—its rows are linearly dependent: $(\mathbf{B} - \mathbf{C}) + (\mathbf{C} - \mathbf{A}) + (\mathbf{A} - \mathbf{B}) = 0$. In fact, the three points *23 31 12* all lie on the 'unit line'

$$\begin{vmatrix} 0 & 1 & -1 \\ -1 & 0 & 1 \end{vmatrix} = [1 \quad 1 \quad 1].$$

That *23* is the harmonic conjugate of **23** with respect to **B** and **C** can be shown algebraically by taking $\theta = Y/Z$ as homographic parameter on the line $\mathbf{AB} = [100]$ (i.e. the line $X = 0$). For these four points this parameter is $-1, 1, \infty$ and 0, respectively, and hence the cross-ratio is -1.

3.8 Configurations in More than Two Dimensions

Four points in 3-space, not all in one plane, determine six lines and four planes. This is a *tetrahedron* (α_3), which can be described as a $(4_3 6_2)$ of points and lines, a $(6_2 4_3)$ of lines and planes, or a (4_3) of points and planes. We need a notation that will summarize this kind of description. We shall say that a tetrahedron is a configuration

4	3	3
2	6	2
3	3	4

The numbers along the diagonal denote the number of points, lines and planes. The off-diagonal numbers in the $(j + 1)$th row denote the number of i-spaces contained in or containing each j-space.

As further simple examples of configurations in 3-space we note that an icosahedron and a dodecahedron can be described as the configurations

12	5	5
2	30	2
3	3	20

20	3	3
2	30	2
5	5	12

of points, lines and planes. Notice how the fact that they are dual to each other shows up in the fact that the two arrays are interchanged by a half-turn. The tetrahedron, of course, is self-dual.

Generalizing the tetrahedron to higher dimensions, a *simplex* (α_N) in N-space is defined by $N + 1$ points not all lying in an $(N - 1)$-space (hyperplane), the lines through pairs of the points, the planes through three of the points, and so on. For example, a 4-simplex, a 5-simplex and a 6-simplex can be described as configurations

5	4	6	4
2	10	3	3
3	3	10	2
4	6	4	5

6	5	10	10	5
2	15	4	6	4
3	3	20	3	3
4	6	4	15	2
5	10	10	5	6

7	6	15	20	15	6
2	21	5	10	10	5
3	3	35	4	6	4
4	6	4	35	3	3
5	10	10	5	21	2
6	15	20	15	6	7

The relation of these symbols to 'Pascal's triangle' is obvious. The numbers are binomial coefficients. The number of j-spaces in an N-dimensional simplex is $\binom{N+1}{j+1}$ and since each of these j-spaces contains a j-dimensional simplex, the number of i-spaces in each j-space is $\binom{j+1}{i+1}$. The number of j-spaces that each i-space is contained in is also $\binom{j+1}{i+1}$.

3.9 Five Points in 3-Space

Projecting a four-dimensional simplex (α_4) on to a 3-space gives a *complete five-point*—a configuration

5	4	6
2	10	3
3	3	10

The homogeneous coordinates of the five vertices can be chosen ('without loss of generality') to satisfy

$$\mathbf{A} + \mathbf{B} + \mathbf{C} + \mathbf{D} + \mathbf{E} = 0.$$

A *diagonal point* of the configuration is the intersection of an edge with the opposite plane (for example the intersection of the edge **AB** with the plane **CDE** is the point $\mathbf{A} + \mathbf{B} = \mathbf{C} + \mathbf{D} + \mathbf{E}$). There are ten of these diagonal points,

$$\begin{array}{cccc} \mathbf{A}+\mathbf{B} & \mathbf{A}+\mathbf{C} & \mathbf{A}+\mathbf{D} & \mathbf{A}+\mathbf{E} \\ \mathbf{B}+\mathbf{C} & \mathbf{B}+\mathbf{D} & \mathbf{B}+\mathbf{E} & \\ \mathbf{C}+\mathbf{D} & \mathbf{C}+\mathbf{E} & & \\ \mathbf{D}+\mathbf{E} & & & \end{array}$$

In an obvious notation, we may call them

$$\begin{array}{cccc} 12 & 13 & 14 & 15 \\ 23 & 24 & 35 & \\ 34 & 35 & & \\ 45 & & & \end{array}$$

3.10 Six Planes in 3-Space

There are ten *harmonic points*

$$
\begin{array}{cccc}
A-B & A-C & A-D & A-E \\
B-C & B-D & B-E & \\
C-D & C-E & & \\
D-E & & &
\end{array}
$$

We may call them

$$
\begin{array}{cccc}
12 & 13 & 14 & 15 \\
23 & 24 & 35 & \\
34 & 35 & & \\
45 & & &
\end{array}
$$

These ten harmonic points are collinear in threes on ten lines and coplanar in sixes on five planes, forming the three-dimensional Desargues' configuration (recall Fig. 1.6):

10	3	3
3	10	2
6	4	5

This is the *dual* of a complete five-point, consisting of five general planes in 3-space and all the line and point intersections. It can be obtained as a *section* of a four-dimensional simplex (α_4) by a general three-space (hyperplane). *Projecting* it on to a plane gives a planar Desargues' (10_3) of points and lines (Fig. 1.7). Alternatively, we could first project to 3-space and then make a plane section—arriving at a planar Desargues' (10_3) by a different route.

3.10 Six Planes in 3-Space

Six general planes in 3-space determine a

20	3	3
4	15	2
10	5	6

obtainable by a three-dimensional *section* of an α_5. We may name the six *hyperplanes* (4-spaces) of an α_5 **123456**. A general 3-space intersects these hyperplanes in six planes, which inherit these same six labels. The 15 lines are the intersections of pairs of these planes and can be denoted by duads; the points are intersections of three planes and can be denoted by triads. These triads label the points of the

Fig. 3.5 A $(20_3 15_4)$

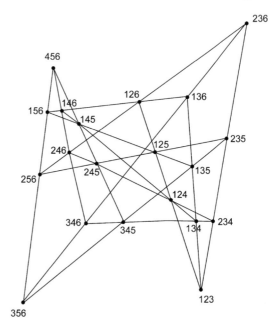

$(20_3 15_4)$ in Fig. 3.5. The duads labeling lines have been omitted—it is easy to see how to insert them. For example, the point **123** is contained in the three lines **12**, **23** and **31**, the line **12** contains the four points **123**, **124**, **125** and **126**, and so on. Figure 3.5 is of course a plane projection of the three-dimensional configuration—a wire model would make things clearer. It is instructive to look for the six planes each containing 10 points and five lines.

It is clear from its derivation from an α_5 that the symmetry group of this configuration is the same as that of α_5, the group S_6 of all permutations of six objects—a group of order $6! = 720$.

This $(20_3 15_4)$ of points and lines is the diagram for the following generalization of Desargues' theorem:

> *Three triangles perspective from a common vertex are perspective from three concurrent axes.*

Three general planes in a 3-space determine a *trihedron*, which consists of the three planes, say, **1**, **2** and **3**, their three lines of intersection **23**, **31** and **12** and their point of intersection **123**. The planes and lines are the *faces* and *edges* of the trihedron and the point is its *vertex*. A trihedron is the 3-space dual of a triangle.

Six general planes in 3-space can be regarded as two sets of three planes, in $\frac{1}{2}\binom{6}{3} = 10$ different ways. The configuration under discussion contains ten associated trihedron pairs. Look, for instance, at the trihedron pair {**123**, **456**} in Fig. 3.5.

3.10 Six Planes in 3-Space

Fig. 3.6 Alternative labeling scheme for the $(20_3 15_4)$

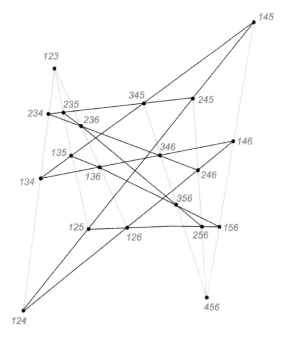

The faces of each trihedron intersect the faces of the other in nine lines

$$\begin{array}{ccc} 14 & 15 & 16 \\ 24 & 25 & 26 \\ 34 & 35 & 36 \end{array}$$

The rows of this array specify three triangles perspective from the vertex **456**. The planes in which they lie are the planes of the trihedron **123**, from which it follows that they are perspective from the edges of the trihedron **123**—the three concurrent axes **12 23 31**. The columns specify three triangles perspective from the vertex **123** and from the three concurrent axes **56 64 45**. Figure 3.5 can be regarded as an illustration of the theorem (or its converse...) in 15 different ways—every point of the figure is a vertex for three perspective triangles.

An alternative labeling scheme arises if we denote the six *vertices* of an α_5 by *123456*, label its edges by duads, its planes by triads and its three-dimensional subspaces by tetrads. These triads and tetrads are then inherited by the lines and points of the three-dimensional section. A general 3-space cuts the planes to give points $(3+2=5+0)$, the 3-spaces to give lines $(3+3=5+1)$. The points of the $(20_3 15_4)$ in 3-space can be labeled by triads, the lines by tetrads. In this notation, for example, the line *1234* contains the four points *123*, *234*, *314* and *124*, and the point *123* lies on the three lines *1234*, *1235* and *1236* (which could be given the obvious simpler alternative names **56**, **64** and **45**). In this scheme the triad denoting one of the 20 points is the complement of the triad labeling the same point in the previous scheme—the label **123** is replaced by **456**, etc. This alternative labeling scheme is shown in Fig. 3.6. The duads denoting the lines are unchanged.

3.11 Six Points in 4-Space

Six general points in 4-space determine a configuration

6	5	10	10
2	15	4	6
3	3	20	3
4	6	4	15

called '*Richmond's hexastigm*'. It is obtainable as projection of an α_5 on to a general 4-space. Its 15 lines and 20 planes are called *Plücker lines* and *Steiner planes*. Let the homogeneous coordinates of the six points be denoted by **A**, **B**, **C**, **D**, **E** and **F**. Since these six points all lie in the same 4-space their coordinate sets are linearly dependent, and since a set of homogeneous coordinates has an overall adjustable factor we can, without loss of generality, take these coordinates to satisfy

$$\mathbf{A} + \mathbf{B} + \mathbf{C} + \mathbf{D} + \mathbf{E} + \mathbf{F} = 0.$$

Fifteen points are defined as the intersections of edges with opposite (three-dimensional) 'faces'. They are

$$\begin{array}{ccccc}
\mathbf{A}+\mathbf{B} & \mathbf{A}+\mathbf{C} & \mathbf{A}+\mathbf{D} & \mathbf{A}+\mathbf{E} & \mathbf{A}+\mathbf{F} \\
 & \mathbf{B}+\mathbf{C} & \mathbf{B}+\mathbf{D} & \mathbf{B}+\mathbf{E} & \mathbf{B}+\mathbf{F} \\
 & & \mathbf{C}+\mathbf{D} & \mathbf{C}+\mathbf{E} & \mathbf{C}+\mathbf{F} \\
 & & & \mathbf{C}+\mathbf{E} & \mathbf{D}+\mathbf{F} \\
 & & & & \mathbf{E}+\mathbf{F}
\end{array}$$

These *Cremona points* can be conveniently denoted by duads:

$$\begin{array}{cccc}
12 & 13 & 14 & 15 & 16 \\
 & 23 & 24 & 25 & 26 \\
 & & 34 & 25 & 36 \\
 & & & 45 & 46 \\
 & & & & 56
\end{array}$$

The 15 *harmonic points*

$$\begin{array}{ccccc}
\mathbf{A}-\mathbf{B} & \mathbf{A}-\mathbf{C} & \mathbf{A}-\mathbf{D} & \mathbf{A}-\mathbf{E} & \mathbf{A}-\mathbf{F} \\
 & \mathbf{B}-\mathbf{C} & \mathbf{B}-\mathbf{D} & \mathbf{B}-\mathbf{E} & \mathbf{B}-\mathbf{F} \\
 & & \mathbf{C}-\mathbf{D} & \mathbf{C}-\mathbf{E} & \mathbf{C}-\mathbf{F} \\
 & & & \mathbf{C}-\mathbf{E} & \mathbf{D}-\mathbf{F} \\
 & & & & \mathbf{E}-\mathbf{F}
\end{array}$$

3.11 Six Points in 4-Space

can be called

$$\begin{array}{ccccc} 12 & 13 & 14 & 15 & 16 \\ 23 & 24 & 25 & 26 & \\ 34 & 25 & 36 & & \\ 45 & 46 & & & \\ 56 & & & & \end{array}$$

The points **12** and *12*, for example, are harmonic conjugates with respect to the point pair **AB**. This can be seen if we take **ABCDE** as the reference simplex (α_4) for the 4-space, so that $\mathbf{A} = (10000)$, $\mathbf{B} = (01000)$, $\mathbf{A} + \mathbf{B} = (11000)$ and $\mathbf{A} - \mathbf{B} = (1 -1\, 000)$. Taking the first two coordinates as homogeneous coordinates for the line **AB**, the homographic parameters of these four points are ∞, 0, 1 and -1.

The 15 harmonic points determine a *dual hexastigm*

$$\begin{array}{|cccc|} \hline 15 & 4 & 6 & 4 \\ 3 & 20 & 3 & 3 \\ 6 & 4 & 15 & 2 \\ 10 & 10 & 5 & 6 \\ \hline \end{array}$$

which is a four-dimensional *section* of an α_5. Its lines, planes and 3-spaces may be called the harmonic points, lines, planes, and 3-spaces. They can be denoted, in an obvious notation, by triads, tetrads and pentads. The three points *23*, *31* and *12*, for example, lie on the line *123*, the four lines *123*, *234*, *314* and *124* lie in the plane *1234*.

Further incidence properties now arise. The 20 *Steiner planes* of the original hexastigm each contain three of the harmonic points. We can readily deduce, for example, that the point *12* is contained in the four planes **ABC**, **ABD**, **ABE** and **ABF**, and the plane **ABC** contains the three points *23*, *31* and *12*. Hence

> *The harmonic points and Steiner planes constitute a* $(15_4 20_3)$ *of points and planes.*

The line *123* is contained in the three 3-spaces **ABCD**, **ABCE**, **ABCF**, and **ABEF** and the 3-space **ABCD** contains the lines *123*, *234*, *314* and *124*. Hence

> *The harmonic lines and the 3-spaces of the original hexastigm constitute a* $(20_3 15_4)$.

The point *12* is contained in **ABCD**, **ABCE**, **ABCF**, **ABDE**, **ABDF** and **ABEF** and the 3-space **ABCD** contains the points *23*, *31*, *12*, *14*, *24* and *34*. Hence

> *The harmonic points and the 3-spaces of the original hexastigm constitute a* (15_6).

This (15_6) is self-dual.

Fig. 3.7 The (15₃) of duads and synthemes

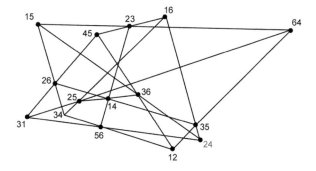

3.12 Sylvester's Duads and Synthemes

The 15 Cremona points, denoted by the duads

$$\begin{array}{ccccc} 12 & 13 & 14 & 15 & 16 \\ & 23 & 24 & 25 & 26 \\ & & 34 & 35 & 36 \\ & & & 45 & 46 \\ & & & & 56 \end{array}$$

are *collinear in threes* because the basic linear dependence can be written in 15 ways such as

$$(\mathbf{A}+\mathbf{B})+(\mathbf{B}+\mathbf{C})+(\mathbf{D}+\mathbf{F}) = 0$$

The 15 *Sylvester lines* on which they lie may be denoted by *synthemes*, which are symbols consisting of three points **12**, **34** and **56**, for example, are contained in the line **12.34.46** and the point **12** is contained in **12.34.56**, **12.35.64** and **12.36.45**. (Note that the order of the two symbols comprising a duad and the order of the three duads comprising a syntheme are of no consequence.)

This configuration (15₃) of Cremona points and Sylvester lines (Fig. 3.7 is a plane projection of it) has some fascinating properties, as we shall see.

Its symmetry group is the group S_6 of all permutations of six things (of order $6! = 720$), because permutations of the six points **ABCDEF** from which it was derived leave it unchanged.

An alternative labeling scheme is possible in which the Cremona points are labeled by *synthemes* and the Sylvester lines by *duads*. This can be done in many

3.12 Sylvester's Duads and Synthemes

Fig. 3.8 A Levi graph for the (15_3) (The Tutte-Coxeter graph)

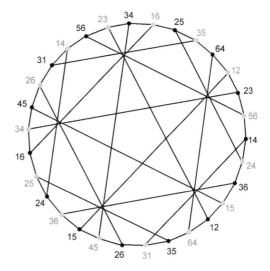

ways, one of which is given by this table:

	1	*2*	*3*	*4*	*5*	*6*
1	–	**23.64.15**	**35.26.14**	**42.31.56**	**54.36.12**	**61.25.34**
2	**23.64.15**	–	**34.56.12**	**41.25.36**	**53.24.16**	**62.45.31**
3	**35.26.14**	**34.56.12**	–	**45.16.23**	**52.31.64**	**63.24.15**
4	**42.31.56**	**41.25.36**	**45.16.23**	–	**51.26.34**	**64.35.12**
5	**54.36.12**	**53.24.16**	**52.31.64**	**51.26.34**	–	**65.14.23**
6	**61.25.34**	**62.45.31**	**63.24.15**	**64.35.12**	**65.14.23**	–

which is interpreted like this: (for example) the line *12* contains the points **23**, **64** and **15** and the point **12** is contained in the lines *23*, *64* and *15*. This particular labeling scheme makes the self-duality of the (15_3) self-evident. The same information is contained in the table obtained from this one by simply interchanging italic and bold fonts.

The incidence properties can also be presented as a Levi graph of the (15_3) (Fig. 3.8). A family of five Sylvester lines can be selected so that no two of them have a common Cremona point (and hence all 15 Cremona points are contained in the family). Each of the columns *1*, *2*, *3*, *4*, *5*, *6* of the table is such a family of Sylvester lines. (The rows of the table, of course, also list the same six families.) Any two families have just one line in common: *3* and *4*, for example, both contain the line **45.61.23** (that is, the line *34*).

The (15_3) is related to the hexastigm **ABCDEF** from which it was constructed in unexpected ways. Observe, for example, that the Silvester line **12.34.56** (*23* in the alternative notation) cuts the Plücker lines **AB**, **CD** and **EF** at **12**, **34** and **56**, respectively, and AB, **12.34.56**, **12.35.46** and **12.36.45** are concurrent at **12**:

> Each Sylvester line intersects three of the Plücker lines and each Plücker line is concurrent with three Sylvester lines.

The join of any pair of Cremona points that is not a Sylvester line is called a *Pascal line*. A Pascal line is determined by two duads that have a common symbol (for example **12** and **13**). There are six possibilities for the common symbol and then $\binom{5}{2} = 10$ possibilities for the remaining two symbols. So there are *sixty* Pascal lines.

> There are eight Pascal lines through every Cremona point.

For example, **12** is joined by Pascal lines to **13**, **14**, **15**, **16**, **23**, **24**, **25** and **26**.

> Every Steiner plane contains three Pascal lines.

For example, the plane **ABC** contains the triangle whose vertices are the Cremona points **23**, **31** and **12**. Its edges are Pascal lines.

Consider now a triangle such as **12 13 45** (with only one repeated symbol in the three duads). One of its edges is a Pascal line and the other two are Sylvester lines. The plane containing this triangle therefore contains two more Cremona points, **36** and **26**. So there are three more Pascal lines in this plane, namely **36 26**, **12 26** and **13 36**. There are 45 planes of this kind, and each Pascal line is contained in three of them. For example, 12 23 is contained in **12 13 45**, **12 13 56** and **12 23 64**. Therefore:

> The sixty Pascal lines belong to a $(60_3 45_4)$ of lines and planes.

3.13 Permutations of Six Things

There are N! permutations of N objects. All these permutations constitute the *symmetric group* S_N. We have been appealing to the permutations of S_6 to deduce general results from particular examples because, of course, all the incidence properties of the configurations we have considered remain valid if we permute the six points ABCDEF of the underlying hexastigm.

A permutation group can act *on itself*. For example, if we apply the permutation (1234) to the transpositions (12), (13) and (123) we get (23), (24) and (234), respectively. Applying any particular permutation to all the permutations of the group, the multiplication table of the group remains valid: in the above example, (12)(13) = (123), which gives (correctly!) (23)(24) = (234). This is fairly obvious—it really amounts to simply renaming the N permuted objects. The action of an element of the permutation group on the whole group in this way is called an *inner automorphism* of the group.

The scheme of duads and synthemes reveals something rather surprising about S_6.

3.14 Another Extension of Desargues' Theorem

Fig. 3.9 A $(15_4 20_3)$

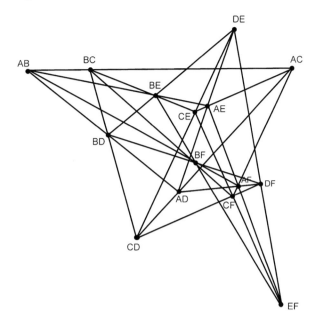

For every permutation of the hexastigm points **ABCDEF** there is a corresponding permutation of the families of Sylvester lines *1, 2, 3, 4, 5* and *6*. The transposition (BC), which we may write as (**23**), for example, corresponds to the product of three disjoint transpositions (**34**)(**65**)(**21**). And, conversely, (**34**)(**65**)(**21**) corresponds to (**23**). We can read off these correspondences from the table. If a and b are any two permutation in S_6 and a' and b' are their images under this mapping, then $a'b' = (ab)'$. We have here an automorphism of S_6 that maps transpositions to products of three disjoint transpositions and vice versa. This obviously cannot be an inner automorphism. It is an *outer automorphism* of S_6.

S_6 is the only symmetric group S_N that has outer automorphisms.

3.14 Another Extension of Desargues' Theorem

Recall that the hexastigm we have been considering is essentially a four-dimensional configuration obtained by projection from an α_5. A *section* of the hexastigm by a 3-space gives a *self-dual* configuration

15	4	6
3	20	3
6	4	15

The $\binom{6}{2} = 15$ points have come from the edges of the α_5 and the $\binom{6}{3} = 20$ lines have come from the planes of the α_5. The points can therefore be labeled by duads from the six symbols **ABCDEF**, and the lines can be denoted by triads. For example, the line **ABC** contains the three points **BC**, **CA** and **AB** and the point **AB** is contained in the four lines **ABC**, **ABD**, **ABE** and **ABF**.

This $(15_4 20_3)$ of points and lines (Fig. 3.9) is the diagram for the following extension of Desargues' theorem:

> *Three triangles perspective in pairs from three collinear vertices are perspective from a common axis.*

For example:

AE BE CE and **AF BF CF** are perspective from the point **EF** and the line **ABC**
AF BF CF and **AD BD CD** are perspective from the point **FD** and the line **ABC**
AD BD CD and **AE BD CD** are perspective from the point **DE** and the line **ABC**.

3.15 Twenty-Seven Lines

A set of 12 lines in 3-space denoted by

$$a_1 \quad a_2 \quad a_3 \quad a_4 \quad a_5 \quad a_6$$
$$b_1 \quad b_2 \quad b_3 \quad b_4 \quad b_5 \quad b_6$$

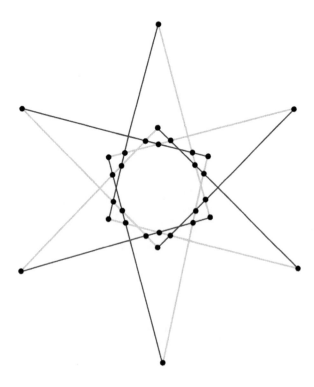

Fig. 3.10 Schläfli's double-six

3.15 Twenty-Seven Lines

such that any two of them intersect if and only if their labels are *not* in the same row or column of this array is called *Schläfli's double-six*. There are 20 points of intersection, so the configuration is a $(30_2 12_5)$ of points and lines (Fig. 3.10 is a two-dimensional indication of this configuration in which a-lines and b-lines are distinguished by grey and black).

Now, since two lines in 3-space that intersect lie in the same plane, there are 30 planes, and we can define 15 more *associated lines* as the intersections of pairs of these planes:

$$c_{12} \quad c_{13} \quad c_{14} \quad c_{15} \quad c_{16}$$
$$c_{23} \quad c_{24} \quad c_{25} \quad c_{26}$$
$$c_{34} \quad c_{35} \quad c_{36}$$
$$c_{45} \quad c_{46}$$
$$c_{56}$$

where, for example, $c_{12} = a_1 b_2 \cdot a_2 b_1$.

Now each of these 27 lines intersects ten others. For example,

a_1 intersects	b_2	b_3	b_4	b_5	b_6	c_{12}	c_{13}	c_{14}	c_{15}	c_{16}
b_1 intersects	a_2	a_3	a_4	a_5	a_6	c_{12}	c_{13}	c_{14}	c_{15}	c_{16}
c_{12} intersects	a_1	a_2	b_1	b_2	c_{34}	c_{35}	c_{36}	c_{45}	c_{46}	c_{56}

and, of course, all the statements got from these three by permuting the indices 123456.

This introduces 105 more points, 30 points such as $a_1 \cdot c_{12}$ (6 choices for the a-line and 5 choices for the c-line), 30 points such as $b_1 \cdot c_{12}$ and 45 points such as $c_{12} \cdot c_{34}$ (15 choices for the first c-line and six choices for the second; divide by 2 because $c_{12} \cdot c_{34}$ is the same as $c_{34} \cdot c_{12}$). The extended configuration is a $(135_2 27_{10})$.

How many planes? There are *forty-five triangles*, each determining a plane: 30 such as $a_1 b_2 c_{12}$ and 15 such as $c_{12} c_{34} c_{56}$. The configuration is a

135	2	3
10	27	5
27	3	45

It is astonishing that this configuration is *regular*. Its symmetries (incidence-preserving permutations of the 27 lines) are transitive on the points, the lines and the planes. All the points are equivalent, as are all the lines and all the planes.

The configuration contains *thirty-six double-sixes*: the original one from which it was constructed, plus *twenty* obtained from

$$\begin{array}{cccccc} c_{23} & c_{31} & c_{12} & b_4 & b_5 & b_6 \\ a_1 & a_2 & a_3 & c_{56} & c_{64} & c_{45} \end{array}$$

by permuting the indices, and *fifteen* like this one:

$$\begin{array}{cccccc} a_1 & b_1 & c_{23} & c_{24} & c_{25} & c_{26} \\ a_2 & b_2 & c_{13} & c_{14} & c_{15} & c_{16} \end{array}$$

Any one of these double-sixes, of course, determines the same configuration. The 12 lines of a double-six can be permuted in $2 \times 6!$ ways (6! permutations of the columns of the array, and transposition of its rows). It follows that the number of symmetries of the configuration—permutations of the 27 lines that preserve incidences—is $2 \times 36 \times 6! = 51840$.

The discovery of this remarkable configuration arose from the study of cubic surfaces. A cubic surface in 3-space is given by an equation $f = 0$ where f is a homogeneous cubic polynomial in the four homogeneous coordinates. A very surprising theorem is

> *A general cubic surface contains* 27 *lines and has* 45 *tritangent planes, constituting a configuration* $(27_5 45_3)$ *of lines and planes, determined by any double-six of the lines.*

A proof will be given later, in the section devoted to cubic surfaces.

3.16 Associated Trihedron Pairs

Three skew (i.e., non-intersecting) lines of the configuration of 27 lines constitute a *triplet*. Every triplet belongs to a *complementary pair* of triplets in which each line of one triplet intersects just two lines of the other. An obvious example is

$$(a_1 \quad a_2 \quad a_3) \quad (b_1 \quad b_2 \quad b_3).$$

A complementary pair of triplets is essentially a skew hexagon—in this case the hexagon $(a_1 b_2 a_3 b_1 a_2 b_3)$. There are 360 such complementary pairs of triplets.

For any chosen double-six, the set of 15 associated lines contains $\binom{6}{3} = 10$ pairs of complementary triplets. Choose one—for example,

$$(c_{23} \quad c_{31} \quad c_{12}) \quad (c_{56} \quad c_{64} \quad c_{45}).$$

The remaining nine associated lines can be written in an array

$$\begin{array}{ccc} c_{14} & c_{25} & c_{36} \\ c_{35} & c_{16} & c_{24} \\ c_{26} & c_{34} & c_{15} \end{array}$$

3.16 Associated Trihedron Pairs

Fig. 3.11 An associated trihedron pair

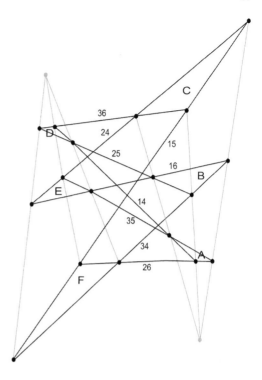

in which no index occurs more than once in a row or in a column. The three lines in a row or a column are edges of a triangle, which lies, of course, in a plane. We have six triangles lying in six planes. We shall call the triangles (or planes) specified by the columns A, B and C, and those specified by the rows D, E and F. The configuration of the nine lines now looks like Fig. 3.11, in which the nine lines are labeled by grey duads and the triangles (or planes) are also identified. *The six grey lines are not among the twenty-seven*. The three on the left are the intersections of pairs of the planes A, B and C and the three on the right are the intersections of D, E and F. We have met this figure before! It is an *associated trihedron pair* (recall Fig. 3.5) that can be interpreted as a diagram of three triangles perspective from a common vertex and from three concurrent axes.

The lines and planes of the configuration of '27 lines and 45 tritangent planes of a cubic surface' is a $(27_5 45_3)$ of lines and planes. Pairs of tritangent planes intersect to give $\binom{45}{2} = 990$ lines. Each line is contained in five of the planes and so each of the lines is an intersection of $\binom{5}{2} = 10$ pairs of the planes. The number of lines of intersection of the 45 planes that do not belong to the set of 27 (i.e., lines of intersections of tritangent planes of a general cubic surface that do not lie in the surface) is therefore $990 - 270 = 720$. Since each associated trihedron pair has six of these lines (the grey lines in Fig. 3.6 for example), the number of trihedron pairs associated with the configuration of the 27 lines is $720/6 = 120$.

3.17 Segre's Notation

The labeling of the 27 lines a_i, b_i and c_{ij} was most convenient for deriving the structure. But once one has established that all the lines are in fact equivalent and that one can begin from any of thirty-six double-sixes, we realize that partitioning the 27 as $12 + 15$ is artificial. However, there *is* a way of labeling the 27 lines in a fully impartial way. Let the symbols μ and ν take the values 1, 2 or 3, and let the 27 triads

$$0\mu\nu \quad \nu 0\mu \quad \mu\nu 0$$

denote 27 lines. Two triads that agree in exactly *one* position correspond to a pair of *intersecting* lines and two triads that agree in *two* positions or *none* correspond to a pair of *skew* lines. This then describes precisely the figure that we have been discussing. For example,

$$\begin{array}{cccccc} 011 & 012 & 013 & 120 & 220 & 320 \\ 021 & 022 & 023 & 110 & 210 & 310 \end{array}$$

is a double-six and its 15 associated lines are

$$\begin{array}{cccc} 033 & 032 & 101 & 201 & 301 \\ 031 & 102 & 202 & 302 \\ & 103 & 203 & 303 \\ & & 330 & 230 \\ & & & 130 \end{array}$$

3.18 The Polytope 2_{21}

A consequence of Segre's labeling scheme is that there is a correspondence between the 27 lines and the 27 *points* in a complex Euclidean 3-space with (*inhomogeneous*) coordinates (X Y Z)

$$\begin{pmatrix} 0 & -\omega^\mu & \omega^\nu \end{pmatrix} \begin{pmatrix} \omega^\nu & 0 & -\omega^\mu \end{pmatrix} \begin{pmatrix} -\omega^\mu & \omega^\nu & 0 \end{pmatrix}$$

where ω is the cube root of unity. These coordinate triples all satisfy

$$X\bar{X} + Y\bar{Y} + Z\bar{Z} = 2$$

and therefore the 27 points $(x_1\ x_2\ x_3\ x_4\ x_5\ x_6)$ in real six-dimensional Euclidean space defined by

$$X = x_1 + ix_2 \qquad Y = x_3 + ix_4 \qquad Z = x_5 + ix_6$$

all lie on a hypersphere

$$x_1^2 + x_2^2 + x_3^2 + x_4^2 + x_5^2 + x_6^2 = 2$$

of radius $\sqrt{2}$. They are the vertices of a *polytope*. Of the $\binom{27}{2} = 351$ edges, 135 have length $\sqrt{6}$ and 216 have length $\sqrt{3}$. The edges of length $\sqrt{6}$ correspond to pairs of intersecting lines in the projective 3-space and edges of length $\sqrt{3}$ correspond to pairs of skew lines. The 45 triangles of side $\sqrt{6}$ contained in the diametral planes of the polytope correspond to the 45 planes of the configuration in the projective 3-space. There are 72 five-dimensional simplexes α_5 of edge length $\sqrt{3}$. The 12 vertices of a pair of diametrically opposite α_5s correspond to a double-six.

3.19 Desmic Systems

A pair of tetrahedra are said to be *desmic* if each edge of one intersects a pair of opposite edges of the other. (In Fig. 3.12 three of these 12 intersection points happen to lie outside the borders of the drawing.)

Now, a plane section of a tetrahedron α_3 is a complete quadrilateral ($6_2 4_3$). The six points are sections of the edges of the α_3 and the four lines are sections of its faces. To satisfy the desmic relationship the triangular face of each tetrahedron must be the *diagonal triangle* of the quadrilateral in which the plane of that face cuts the other tetrahedron (Fig. 3.13). It follows from the properties of a quadrangle and its diagonal triangle that, if A and B denote two vertices of a tetrahedron of a desmic pair and C and D denote the two intersections of AB with edges of the other tetrahedron, then C and D are *harmonic conjugates* of each other with respect to the point pair AB.

We get a neat canonical form for the coordinates of the eight vertices if we choose one of the tetrahedra ABCD to be the *unit tetrahedron*, so that the coordinates of its

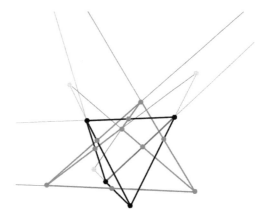

Fig. 3.12 A pair of desmic tetrahedra

Fig. 3.13 Section of one of the tetrahedra (*black*) by a face of the other (*grey*)

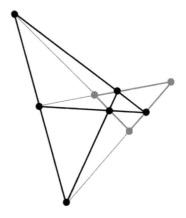

vertices are (1 0 0 0), (0 1 0 0), (0 0 1 0) and (0 0 0 1). We are free to choose a face of the other tetrahedron to be the *unit plane* [1 1 1 1]. Referring to Fig. 3.13, let the grey triangle be the face [0 0 0 1] of the unit tetrahedron and one of the black lines the intersection of this plane with the unit plane. The three points on this line are then (0 1 −1 0), (−1 0 1 0) and (1 −1 0 0). The harmonic conjugate of (0 1 −1 0) with respect to (0 1 0 0) and (0 0 1 0) is (0 1 1 0). *Et cetera*. In this way we get the coordinates of all six points of intersection. One of the grey lines contains the three collinear points (1 −1 0 0), (1 0 1 0) and (0 1 1 0). A plane through this line (other than [0 0 0 1]) must have the form [1 1 −1 α]. Proceeding in this way, we deduce that the planes of the grey tetrahedron are given by the rows of a matrix of the form

$$\begin{bmatrix} 1 & 1 & 1 & 1 \\ 1 & 1 & -1 & \alpha \\ 1 & -1 & 1 & \beta \\ 1 & -1 & -1 & \gamma \end{bmatrix}$$

The same argument for a different face of ABCD reveals that $\alpha = \beta = -\gamma = -1$. The *vertices* of the grey tetrahedron are given by the inverse of this matrix. We arrive at a canonical form for the vertices of any pair of desmic tetrahedra: the columns of

$$\begin{pmatrix} 1 & 0 & 0 & 0 \\ 0 & 1 & 0 & 0 \\ 0 & 0 & 1 & 0 \\ 0 & 0 & 0 & 1 \end{pmatrix} \text{ and } \begin{pmatrix} 1 & 1 & 1 & 1 \\ 1 & 1 & -1 & -1 \\ 1 & -1 & 1 & -1 \\ 1 & -1 & -1 & 1 \end{pmatrix}.$$

3.19 Desmic Systems

Fig. 3.14 Euclidean specialization of a desmic system

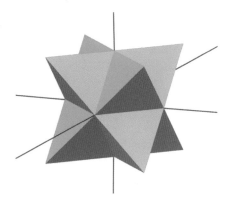

It is easily verified that the two tetrahedrons are perspective from any vertex of the tetrahedron

$$\begin{pmatrix} 1 & 1 & 1 & -1 \\ 1 & 1 & -1 & 1 \\ 1 & -1 & 1 & 1 \\ -1 & 1 & 1 & 1 \end{pmatrix}.$$

Moreover, any pair of tetrahedra from these three is desmic, and the two are perspective from each of the four vertices of the third tetrahedron. Amazing!

A set of three tetrahedra in this relationship is called a *desmic system*.

From the way we have derived the above three matrices, it follows that they provide a canonical form for *any* desmic system: any desmic system can be brought to this form by a collineation:

All desmic systems are projectively equivalent.

A desmic system is, admittedly, not easy to visualize. But if we choose [0001] as the *plane at infinity* we can get a simple affine specialization of the system. If the homogeneous coordinates of the projective 3-space are $(X\ Y\ Z\ W)$, this amounts to taking $(X\ Y\ Z) = (X/W\ Y/W\ Z/W)$ as (inhomogeneous) coordinates in affine space, or even Cartesian coordinates in Euclidean space. The vertices of the reference tetrahedron become the points at infinity on the X, Y and Z axes, and the origin. The vertices of the second and third tetrahedra have coordinates given by the columns of

$$\begin{pmatrix} 1 & -1 & -1 & 1 \\ 1 & -1 & 1 & -1 \\ 1 & 1 & -1 & -1 \end{pmatrix} \text{ and } \begin{pmatrix} -1 & 1 & 1 & -1 \\ -1 & 1 & -1 & 1 \\ -1 & -1 & 1 & 1 \end{pmatrix}.$$

The desmic system then looks like Fig. 3.14.

There is more! The 12 points of intersection of edges and the 12 planes containing pairs of intersecting edges form another desmic system consisting of the three tetrahedra

$$\begin{pmatrix} 0 & 1 & 0 & 1 \\ 1 & 0 & 1 & 0 \\ 0 & -1 & 0 & 1 \\ -1 & 0 & 1 & 0 \end{pmatrix} \begin{pmatrix} 0 & 1 & 0 & 1 \\ 0 & -1 & 0 & 1 \\ 1 & 0 & 1 & 0 \\ -1 & 0 & 1 & 0 \end{pmatrix} \begin{pmatrix} 1 & 0 & 1 & 0 \\ 0 & 1 & 0 & 1 \\ 0 & 1 & 0 & -1 \\ 1 & 0 & -1 & 0 \end{pmatrix}$$

These three have the same 18 edges as the original three. We have an *associated pair of desmic systems*.

An associated pair of desmic systems is a self-dual configuration

24	3	9
4	18	4
9	3	24

It is regular, with a symmetry group of order $2^8 3^3 = 6912$, realizable as a group of collineations.

The perspectivities of a desmic system imply the existence of 16 lines containing the tetrahedron vertices in threes—a $(12_4 16_3)$ configuration—and 16 lines through each of which pass three faces. These 32 lines are the same for both desmic systems of an associated pair, but the roles of the two sets of 16 are reversed for the two systems.

3.20 Baker's Configuration

Returning now to the configuration (15_3) of *Cremona points* and *Sylvester lines* derived from six general real points in 4-space, let us consider the *conjugate complex pair of equianharmonic points* determined by any three of the Cremona points. For example, for the three collinear Cremona points $\mathbf{A} + \mathbf{B}$, $\mathbf{C} + \mathbf{D}$ and $\mathbf{E} + \mathbf{F}$ the equianharmonic points are

$$(\mathbf{A} + \mathbf{B}) + \omega(\mathbf{C} + \mathbf{D}) + \omega^2(\mathbf{E} + \mathbf{F})$$

and

$$(\mathbf{A} + \mathbf{B}) + \omega^2(\mathbf{C} + \mathbf{D}) + \omega(\mathbf{E} + \mathbf{F}).$$

The properties of ω ensure that these two points are unchanged if we make an even permutation of the three Cremona points, and are interchanged if we make an odd permutation. A conjugate complex pair of equianharmonic points can therefore be

3.20 Baker's Configuration

denoted by a pair of synthemes; the pair in this example are then **12.34.56** and **12.56.34**.

In this new context, where the synthemes name points rather than lines, two synthemes related by an *odd* permutation of the three duads refer to conjugate complex points. There are *thirty* equianharmonic points. It is helpful to display the table of synthemes again here, to emphasize this change of meaning:

	1	2	3	4	5	6	
1	–	23.64.15	35.26.14	42.31.56	54.36.12	61.25.34	a_1
2	23.15.64	–	34.56.12	41.25.36	53.24.16	62.45.31	a_2
3	35.14.26	34.12.56	–	45.16.23	52.31.64	63.24.15	a_3
4	42.56.31	41.36.25	45.23.16	–	51.26.34	64.35.12	a_4
5	54.12.36	53.16.24	52.64.31	51.34.26	–	65.14.23	a_5
6	61.34.25	62.31.45	63.15.24	64.12.35	65.23.14	–	a_6
	b_1	b_2	b_3	b_4	b_5	b_6	

Each of these 30 points can also be denoted by a *duad*, for example **23.64.15** is **12** and **15.64.23** is its complex conjugate **21**.

Each *row* of this table specifies a set of five points in the 4-space—the vertices of a simplex α_4. Call these six simplexes $a_1, a_2, \ldots, a_6$. Similarly, each *column* of the table specifies a simplex. Call these six simplexes $b_1, b_2, \ldots, b_6$. A pair of these 12 simplexes share a common vertex if and only if they do not occur in the same row or the same column of the array

$$a_1 \quad a_2 \quad a_3 \quad a_4 \quad a_5 \quad a_6$$
$$b_1 \quad b_2 \quad b_3 \quad b_4 \quad b_5 \quad b_6$$

– a *double-six* of simplexes!

The symbols labeling the rows and columns give alternative names for the equianharmonic points. For example, **12** is the point **23.64.15**, which is the complex conjugate of the point **21**, which is **15.64.23**. In this notation the vertices of the simplex a_1 are **12 13 14 15 16**, those of b_1 are **21 31 41 51 61**, and so on.

Combining the equianharmonic points and the 15 harmonic points we have *forty-five* points. Following Baker, who discovered all this, we call them *nodes*. We have already introduced the duad notation for the harmonic points, *12* denoting **A** − **B** et cetera. We can also denote them by synthemes. We use the same table for translation between duad and syntheme notation for harmonic points, but in this version the order of symbols in a duad and duads in a syntheme is not relevant:

	1	2	3	4	5	6
1	–	23.64.15	35.26.14	42.31.56	54.36.12	61.25.34
2	23.64.15	–	34.56.12	41.25.36	53.24.16	62.45.31
3	35.26.14	34.56.12	–	45.16.23	52.31.64	63.24.15
4	42.31.56	41.25.36	45.16.23	–	51.26.34	64.35.12
5	54.36.12	53.24.16	52.31.64	51.26.34	–	65.14.23
6	61.25.34	62.45.31	63.24.15	64.35.12	65.14.23	–

We can introduce 15 more simplexes

$$c_{12} \quad c_{13} \quad c_{14} \quad c_{15} \quad c_{16}$$
$$c_{23} \quad c_{24} \quad c_{25} \quad c_{26}$$
$$c_{34} \quad c_{35} \quad c_{36}$$
$$c_{45} \quad c_{46}$$
$$c_{56}$$

The simplex c_{12} is defined by its vertices:

$$\mathbf{12} \quad \mathbf{21} \quad 12.34.56 \quad 12.35.64 \quad 12.36.45$$

and all the other c-simplexes are defined similarly. By searching the table we see that the vertices of c_{12} can also be designated as

$$\mathbf{12} \quad \mathbf{21} \quad 23 \quad 46 \quad 15$$

You can now verify (if you want to!) that the pattern corresponds to the $(27_5 45_3)$ of lines and planes associated with the general cubic surface. We have 27 simplexes and 45 nodes, each node being shared by three of the simplexes. The node **12** for example belongs to a_1, b_2 and c_{12} and *12* (same as *23.64.15*) belongs to c_{23}, c_{64} and c_{15}.

To proceed further, we set up a system of homogeneous coordinates for the 4-space in which these 27 simplexes lie. Let this 4-space be the *unit hyperplane* in a 5-space, so that the points in the 4-space have *six* coordinates satisfying

$$x^1 + x^2 + x^3 + x^4 + x^5 + x^6 = 0.$$

The six general points **A**, **B**, **C**, **D**, **E** and **F** of the hexastigm in the 4-space from which the 27 simplexes have been derived can be taken, without any loss of generality, to be the perspective projection from the unit point of the 5-space, of the vertices of the reference simplex α_5. Then **A**, **B**, **C**, **D**, **E** and **F** are given by the six permutations of $(-5\ 1\ 1\ 1\ 1\ 1)$ (which satisfy $\mathbf{A} + \mathbf{B} + \mathbf{C} + \mathbf{D} + \mathbf{E} + \mathbf{F} = 0$). The

3.20 Baker's Configuration

15 Cremona points are now given by the permutations of $(-2\ -2\ 1\ 1\ 1\ 1)$, the harmonic points by the 15 permutations of $(1\ -1\ 0\ 0\ 0\ 0)$ and the 30 equianharmonic points by the permutations of $(1\ 1\ \omega\ \omega\ \bar{\omega}\ \bar{\omega})$. (For example, $(-2\ 1\ -2\ 1\ 1\ 1)$ is $\mathbf{A} + \mathbf{C}$, $(0\ 0\ 1\ 0\ -1\ 0)$ is 35 and $(1\ \omega\ \bar{\omega}\ \bar{\omega}\ 1\ \omega)$ is $\mathbf{15.26.34} = \mathbf{45}$.)

There are 45 hyperplanes (4-spaces) in the 5-space given by the permutations of $[1\ -1\ 0\ 0\ 0\ 0]$ and $[1\ 1\ \omega\ \omega\ \bar{\omega}\ \bar{\omega}]$. They all contain the unit point. We shall employ these same coordinates to denote the section of these hyperplanes by our 4-space. We thus have a configuration of 45 hyperplanes (3-spaces) in our 4-space that is *dual* to the configuration of 45 nodes. We use square brackets to distinguish the hyperplanes from their corresponding nodes. (For example, $[0\ 0\ 1\ 0\ -1\ 0]$ is $[35]$ and $[1\ \omega\ \bar{\omega}\ \bar{\omega}\ 1\ \omega]$ is $[\mathbf{15.26.34}] = [\mathbf{45}]$.)

Each of the 45 hyperplanes contains 12 of the nodes and each of the 45 nodes is contained on 12 of the hyperplanes.

We have a self-dual (45_{12}) of points and 3-spaces. This is easily verified:

$$[23] = [\mathbf{12.34.56}] = [1\ 1\ \omega\ \omega\ \bar{\omega}\ \bar{\omega}]\text{ contains the 12 nodes}$$

$(1\ \bar{\omega}\ \omega\ 1\ \omega\ \bar{\omega})\quad (1\ \omega\ \bar{\omega}\ \omega\ \bar{\omega}\ 1)\quad (1\ \bar{\omega}\ 1\ \omega\ \omega\ \bar{\omega})\quad (1\ \omega\ \omega\ \bar{\omega}\ 1\ \bar{\omega})$

$(1\ \omega\ \omega\ \bar{\omega}\ \bar{\omega}\ 1)\quad (1\ \bar{\omega}\ 1\ \omega\ \bar{\omega}\ \omega)\quad (1\ \omega\ \bar{\omega}\ \omega\ 1\ \bar{\omega})\quad (1\ \bar{\omega}\ \omega\ 1\ \omega\ \bar{\omega})$

$(1\ 1\ \omega\ \omega\ \bar{\omega}\ \bar{\omega})\quad (1\ -1\ 0\ 0\ 0\ 0)\quad (0\ 0\ 1\ -1\ 0\ 0)\quad (0\ 0\ 0\ 0\ 1\ -1)$

which are

	42	52	62	12
	34	35	36	31
	23	23.14.56	23.15.64	23.16.45

Similarly, $[12]$ contains the 12 nodes

12.34.56	12.56.34	34	56
12.35.46	12.46.35	35	64
12.45.36	12.36.45	36	45

That completes the proof. Duality and S_6 permutations ensure the rest.

We have seen that each of the 45 nodes is a vertex for three of the 27 simplexes. The nodes contained in $[23]$ are vertices of the three simplexes that share the vertex $\mathbf{32}$ (other than $\mathbf{32}$ itself). They are the simplexes b_2, a_3 and c_{23}. Similarly, the simplexes c_{23}, c_{64} and c_{15} share vertex 12 and face $[12]$. That is:

Three simplexes sharing a vertex also share the (three-dimensional) face opposite that vertex.

Since no two of the simplexes share more than one vertex, they share no common edge and no common plane face. The number of lines in the configuration is therefore $27 \times \binom{5}{2} = 270$ and the number of planes is $27 \times \binom{5}{2} = 270$. One can deduce that the 45 nodes belong to a configuration

45	12	18	12
2	270	3	3
3	3	270	2
12	18	12	45

It is regular and self-dual. We call the configuration 'Baker's configuration' because H F Baker investigated it thoroughly and wrote *A Locus with 25920 Linear Self-Transformations*. (The 'locus' is a hypersurface in 4-space called Burkhardt's primal; the nodes are its singular points.)

The S_6 that permutes the coordinates $(x_1 x_2 x_3 x_4 x_5 x_6)$ is a group of projective transformations (collineations) of the 4-space, that permutes the 45 nodes in an obvious way. Employing the properties of ω it is not difficult to deduce that the collineation with matrix

$$\begin{pmatrix} 1 & \omega & \omega & & & \\ \omega & 1 & \omega & & & \\ \omega & \omega & 1 & & & \\ & & & -1 & -\vec{\omega} & -\vec{\omega} \\ & & & -\vec{\omega} & -1 & -\vec{\omega} \\ & & & -\vec{\omega} & -\vec{\omega} & -1 \end{pmatrix}$$

leaves the 4-space [111111] unchanged and produces the permutation

(*14* 14.23.56 14.56.23)(*15* 15.23.64 15.64.23)(*16* 16.23.45 16.45.23)
(*24* 24.31.56 24.56.31)(*25* 25.31.64 25.64.31)(*26* 26.31.45 26.45.31)
(*34* 34.12.56 34.56.12)(*35* 35.12.64 35.64.12)(*36* 36.12.45 36.45.12)
(14.25.36 15.26.34 16.24.35)(15.36.24 16.34.25 14.35.26)
(41.62.35 16.25.34 15.24.36)(16.35.24 15.34.26 14.36.25)
(23)(31)(12)(56)(64)(45)

of the nodes. Or, in the briefer notation,

(*14* 56 65)(*15* 12 21)(*16* 34 43)
(*24* 14 41)(*25* 35 53)(*26* 62 26)
(*34* 32 23)(*35* 46 64)(*36* 15 51)
(24 45 52)(36 61 13)(31 16 63)(25 54 42)
(23)(31)(12)(56)(64)(45).

3.20 Baker's Configuration

This transformation and the S_6 together generate a group of symmetries of the configuration, of order $36 \times 6! = 25920$. The distinction between harmonic and equianharmonic nodes disappears if we allow complex projective transformations. All the 45 nodes are equivalent. The full symmetry group of order $2 \times 36 \times 6! = 51840$ includes complex conjugation.

Since three of the simplexes that share a vertex also share the hyperplanar 'face' opposite that vertex, the 12 nodes in each hyperplane are the vertices of three tetrahedra. For example, as we have seen, the hyperplane [*12*] contains the 12 nodes

$$\begin{array}{cccc} \mathbf{32} & \mathbf{23} & 34 & 56 \\ \mathbf{64} & \mathbf{46} & 35 & 64 \\ \mathbf{51} & \mathbf{15} & 36 & 45 \end{array}$$

Each row specifies a *tetrahedron*. It is easy to check that each edge of one tetrahedron meets a pair of opposite edges of the other.

> *The three tetrahedra in each hyperplane of Baker's configuration constitute a desmic system.*

Now, since in a desmic system each pair of tetrahedra is perspective from each of the four vertices of the third tetrahedron, there are in each of the hyperplanes 12 more lines, not belonging to the configuration, each containing three nodes. These are the κ-lines. There are *twenty* such as

$$23 \quad 31 \quad 12$$

forty such as

$$\mathbf{14.25.36} \quad \mathbf{15.26.34} \quad \mathbf{16.24.35}$$

and a *hundred and eighty* such as

$$12 \quad \mathbf{13.24.56} \quad \mathbf{23.14.56}$$

The 240 κ-lines are all equivalent under the symmetries of the configuration.

Chapter 4
Quadratic Figures

Abstract Canonical forms for conics are given, and tangents and 'polarity' are introduced. Various metric planes are derived from the projective plane by specializing an 'absolute conic'. Pascal's *hexagrammum mysticum* and the extended Pascal configuration of 60 Pascal lines are discussed. Quadric surfaces are introduced and their affine classification is discussed. Various kinds of metric space are derived from projective 3-space containing a specialized 'absolute quadric'.

4.1 Conics

Except for a small digression concerning the complex projective line we have till now been dealing exclusively with *linear* expressions. The next step is to start thinking about homogeneous *quadratic* expressions. In the projective plane, a *conic* is a curve whose points $(X\ Y\ Z)$ satisfy a homogeneous quadratic equation,

$$aX^2 + bY^2 + cZ^2 + 2(dYZ + eZX + fXY) = 0.$$

> A unique conic exists through any five given general points of which no three are collinear.

This is easily proved: if the five given points are $(X_1\ Y_1\ Z_1), \ldots, (X_5\ Y_5\ Z_5)$ the equation of the required conic is

$$\begin{vmatrix} X^2 & Y^2 & Z^2 & YZ & ZX & XY \\ X_1^2 & Y_1^2 & Z_1^2 & Y_1Z_1 & Z_1X_1 & X_1Y_1 \\ X_2^2 & Y_2^2 & Z_2^2 & Y_2Z_2 & Z_2X_2 & X_2Y_2 \\ X_3^2 & Y_3^2 & Z_3^2 & Y_3Z_3 & Z_3X_3 & X_3Y_3 \\ X_4^2 & Y_4^2 & Z_4^2 & Y_4Z_4 & Z_4X_4 & X_4Y_4 \\ X_5^2 & Y_5^2 & Z_5^2 & Y_5Z_5 & Z_5X_5 & X_5Y_5 \end{vmatrix} = 0.$$

We shall adopt a very concise matrix notation, in which the set of coordinates $(X\ Y\ Z)$ of a point A, written as a *column*, will be denoted by A. When written as a *row*, it will be denoted by A^T. Similarly, the set of components $[l\ m\ n]$ of a line, written as a *row*, will be denoted by L and when written as a *column*, by L^T. The above equation for a conic is then simply

$$A^T Q A = 0,$$

where Q is the symmetric matrix

$$\begin{pmatrix} a & f & e \\ f & b & d \\ e & d & c \end{pmatrix}.$$

We shall consider only *real* conics throughout; the coefficients a, b, c, d, e and f are real numbers. But we will allow homogeneous coordinates of points and lines to be complex.

A conic is *degenerate* if its equation factorizes:

$$(l_1 X + m_1 Y + n_1 Z)(l_2 X + m_2 Y + n_2 Z) = 0.$$

A degenerate conic is a pair of lines. The matrix Q of a degenerate conic has the form

$$\begin{pmatrix} l_1 l_2 & (l_1 m_2 + m_1 l_2)/2 & (l_1 n_2 + n_1 l_2)/2 \\ (m_1 l_2 + l_1 m_2)/2 & m_1 m_2 & (m_1 n_2 + n_1 m_2)/2 \\ (n_1 l_2 + l_1 n_2)/2 & (n_1 m_2 + m_1 n_2)/2 & n_1 n_2 \end{pmatrix}.$$

The matrix of a degenerate conic is singular. This could be shown by tediously working out the determinant of this matrix for the general case of a degenerate conic, but that is not necessary. No generality is lost by choosing the two lines to be two reference lines such as [001] and [010] and the result is then obvious. Conversely, a conic is degenerate if its matrix is singular. *Proof*: a singular matrix has a zero eigenvalue, so if Q is singular there is a point A satisfying $QA = 0$. It obviously lies on the conic ($A^T Q A = 0$). Let B be any other point on the conic; $B^T Q B = 0$. Then

$$(A + \lambda B)^T Q (A + \lambda B) = \lambda^2 (B^T Q B) + 2\lambda (B^T Q A) + (A^T Q A) = 0.$$

Hence every point $A + \lambda B$ lies on the conic, for any λ. The conic contains the line AB; the conic is degenerate.

A conic is degenerate if and only if its matrix is singular.

A real line AB cuts a non-degenerate conic at the points $A + \lambda B$ given by the roots of

$$\lambda^2 (B^T Q B) + 2\lambda (B^T Q A) + (A^T Q A) = 0.$$

The two points are real, real but coincident, or a pair of conjugate complex points, according to whether the discriminant

$$(B^T Q A)^2 - (B^T Q B)(A^T Q A)$$

4.2 Tangents

Fig. 4.1 A hyperbola, a parabola and an ellipse in relation to the line at infinity (the *grey line*)

Fig. 4.2 Conic sections

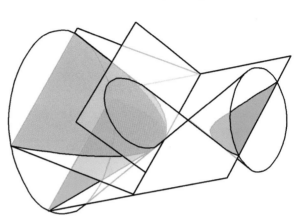

is positive, zero or negative. In the case of coincidence of the roots, the line is a *tangent* to the conic, which *touches* it at $A + \lambda B$.

The familiar *affine classification of conics* arises by choosing a *line at infinity*. A real non-degenerate conic is a *hyperbola*, a *parabola* or an *ellipse*, according to whether it intersects the line at infinity in two real points, touches the line at infinity, or intersects the line at infinity at a pair of conjugate complex points (Fig. 4.1). The *asymptotes* of a hyperbola are the tangents to it that touch it at infinity. In the projective plane there is no such distinction—all real non-degenerate conics are *projectively equivalent*. It is instructive to recall the way homogeneous coordinates were introduced by projection from an affine space on to a 'picture plane' and to relate this to the definition of the 'conic sections' as sections of a cone (Fig. 4.2).

4.2 Tangents

A line $L = [l\ m\ n]$ is a tangent to a non-degenerate conic Q with equation $P^T Q P = 0$ if and only if

$$L Q^{-1} L^T = 0.$$

Proof For any line satisfying this equation, the point $A = Q^{-1} L^T$ lies on the line (because $LA = 0$) and also on the conic (because $(LQ^{-1})Q(Q^{-1}L^T) = 0$). Let B be any other point on the line, not on the conic, so that $LB = 0$ but $B^T Q B \neq 0$. For a point $A + \lambda B$ on the line to lie on the conic, λ must satisfy

$$0 = \left(Q^{-1} L^T + \lambda B\right)^T Q \left(Q^{-1} L^T + \lambda B\right).$$

Q is symmetric ($Q^T = Q$), so this is just

$$0 = \lambda(B^T L^T + LB) + \lambda^2(B^T QB).$$

But $LB = 0$. Hence $\lambda = 0$; A is the only point that lies on the conic and on the line; the line is a tangent to Q.

Conversely, if L is a tangent to Q, A the point of contact and B some other point on the line, we have $LA = LB = A^T QA = 0$. The condition for tangency is that $\lambda = 0$ should be the only solution of

$$0 = (A + \lambda B)^T Q(A + \lambda B) = \lambda(B^T QA + A^T QB) + \lambda^2(B^T QB).$$

That implies $B^T QA = 0$. Therefore B lies on the line $A^T Q$. A also lies on the line $A^T Q$ (because $A^T QA = 0$). Therefore $L = A^T Q$ and so $LQ^{-1}L^T = A^T QQ^{-1}QA = A^T QA = 0$.

The tangent to Q, touching it at A, is $L = A^T Q$.

Similarly (and dually),

The point of contact of a tangent L to Q is $A = Q^{-1} L^T$.

The *dual* of a conic is a *conic envelope*—the set of all tangents to a conic. □

4.3 Canonical Forms

There are several convenient *canonical forms* for the equation of a real non-degenerate conic, depending on the choice of reference system for the homogeneous coordinates. In matrix notation the equation of a conic is

$$A^T QA = 0$$

where P is the column of homogeneous point coordinates and P^T is its transpose. Changing the reference triangle has the effect

$$A \to SA \qquad A^T \to A^T S^T$$

where S is a non-singular matrix and S^T is its transpose. The equation of the conic remains valid in the new system if

$$Q \to S^{-T} Q S^{-1}.$$

A fundamental theorem in matrix theory says that any symmetric matrix Q can be transformed to diagonal form

$$\begin{pmatrix} \mu & & \\ & \nu & \\ & & \rho \end{pmatrix}$$

4.3 Canonical Forms

by a transformation of this kind. μ, ν and ρ are the eigenvalues of Q. None of them is zero because Q is non-singular. A further transformation with

$$S = \begin{pmatrix} \sqrt{|\mu|} & & \\ & \sqrt{|\nu|} & \\ & & \sqrt{|\rho|} \end{pmatrix}$$

($|\mu|$, $|\nu|$ and $|\rho|$ denote absolute values) will not change the reference triangle; it corresponds to a new choice for the unit point. Finally, note that the overall sign of Q has no significance and that permuting X, Y and Z just maps the reference triangle to itself. We arrive at just two possible canonical forms for any real non-degenerate conic:

$$X^2 + Y^2 + Z^2 = 0 \quad \text{and} \quad X^2 + Y^2 - Z^2 = 0.$$

In the first case we have a *virtual conic* that contains no real points at all. (In the *complex* projective plane P(2, C), all non-degenerate conics can be given this form.) The second form can be taken as a *standard canonical form for any real non-degenerate conic* in P(2, R).

The equation of a conic Q acquires a different simple canonical form if we choose the vertices of the reference triangle and the unit point to be four points on the conic:

$$dYZ + eZX + fXY = 0, \qquad d + e + f = 0.$$

This particular canonical form provides a simple proof that

There is a unique non-degenerate conic through any five general points.

(By 'general' points we mean that no three are collinear.) Four of the given points can be chosen as reference points and unit point, so that the equation of the conic has the above form. If the fifth point is $(\mu \nu \rho)$ we have two equations to determine the ratios $d : e : f$,

$$d\nu\rho + e\rho\mu + f\mu\nu = d + e + f.$$

The unique solution is

$$d : e : f = \mu(\nu - \rho) : \nu(\rho - \mu) : \rho(\mu - \nu).$$

Another useful trick is to let two of the vertices of the reference triangle be the points of contact of the tangents through the third vertex (not on the conic), and to let any other point on the conic be the unit point, as in Fig. 4.3.

The equation of the conic becomes

$$Y^2 = ZX$$

which gives a *parametric form* for the equation of the conic:

$$(X\ Y\ Z) = (\theta^2\ \theta\ 1).$$

Fig. 4.3 Reference triangle chosen in relation to a conic

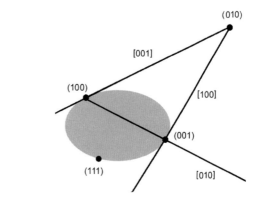

Fig. 4.4 Projecting a line on to a conic

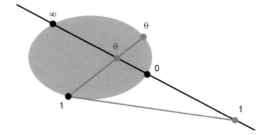

Incidentally, since any non-singular conic can be brought to this form by a collineation, it follows that

> *All non-singular conics are projectively equivalent.*

The parameter θ is a *homographic parameter* on the conic, which makes it possible to regard the conic as a *one-dimensional projective space*. In fact, a projective line in the plane can be projected on to a conic by a *perspectivity* with vertex on the conic, so that the homographic parameters match. This is indicated in Fig. 4.4, where (without loss of generality) the line is chosen to be [010] and the vertex of the perspectivity is (111). The point on the conic marked θ is the general point $(\theta^2\ \theta\ 1)$. The points (100) and (001) have parameters $\theta = \infty$ and $\theta = 0$. The point marked θ *on the line* is $(\theta\ 0\ 1)$ (because $(\theta^2\ \theta\ 1)$, $(\theta\ 0\ 1)$ and $(1\ 1\ 1)$ are collinear). Hence the perspectivity projects the point on the line with homographic parameter $\theta = -X/Z$ to the point on the conic with the same parameter θ. The tangent to the conic at (111) is $[-1\ 2\ -1]$, which intersects [010] at $\begin{vmatrix} -1 & 2 & -1 \\ 1 & 1 & 1 \end{vmatrix} = (1\ 0\ -1)$, which is the point $\theta = 1$ on the line [101].

4.4 Polarity

From now on, the word 'conic' will be taken to mean a *real, non-degenerate* conic.

4.4 Polarity

Fig. 4.5 Pole and polar

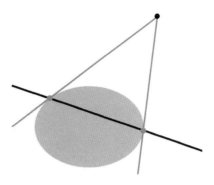

Any conic Q establishes a one-to-one correspondence between the points and the lines of the projective plane that preserves incidences. This *polarity with respect to a conic* therefore converts any figure in the plane to a *dual* figure.

The *pole* of a line L with respect to a conic is the intersection of the tangents through the two points of intersection of the L with Q.

The *polar* of a point A with respect to a conic Q is the line joining the contact points of the two tangents to Q through A.

These two concepts are mutually dual. Observe that it follows immediately from these definitions that if A is the polar of L, then L is the polar of A (and vice versa). The polar of a point *on* the conic is the tangent at that point and the pole of a *tangent* is its point of contact.

Figure 4.5 illustrates the case where the tangents and contact points in these statements are real and distinct. Even when they are not (when the polar is entirely 'outside' the conic and the pole is 'inside'...), these statements are still true. We have already seen that when a line and a conic have no real common points, they have a common pair of conjugate complex points; dually, if a point has no real tangents to a conic, it has a pair of conjugate complex tangents. It is easily shown that the line through a pair of conjugate complex points is *real*, and that the point of intersection of a conjugate complex pair of lines is *real*.

> *The pole of a line* L *with respect to* Q *is the point* $Q^{-1}L^T$ *and the polar of a point* A *with respect to a conic* Q *is the line* A^TQ.

Proof: A line L containing A and B satisfies $LA = LB = 0$. If it cuts the conic Q at A and B we also have $A^TQA = B^TQB = 0$. The tangents to Q at A and at B are respectively A^TQ and B^TQ. They intersect at P, satisfying $A^TQP = B^TQP = 0$. But $A^TQ(Q^{-1}L^T) = B^TQ(Q^{-1}L^T) = 0$. Hence $P = Q^{-1}L^T$.

Polarity is an *incidence-preserving* correspondence between the points and the lines of the projective plane. In other words:

> *If a point* A *lies on the polar of* B *then* B *lies on the polar of* A.

(Figure 4.6). The proof is obvious: because the matrix Q is symmetric, $A^TQB = B^TQA$.

> *The intersection of the polars of two points is the pole of the line joining them.*

Fig. 4.6 Reciprocity of polarity

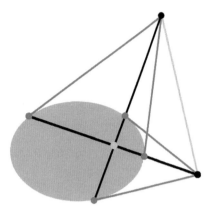

Dually, we have:

> The line through the poles of two lines is the polar of their point of intersection.

Proof The polars of two points A and B are A^TQ and B^TQ. The line L through A and B is given by $LA = LB = 0$. The pole of L is $P = Q^{-1}L^T$. The intersection P' of the two polars is given by $A^TQP' = B^TQP' = 0$. But since $A^TQP = B^TQP = 0$. Therefore $P' = P$. The dual statement can be proved similarly, but that is unnecessary.

4.5 Self-Polar Triangles

A triangle is *self-polar* with respect to a conic if each vertex is the pole of the opposite side. There is one in Fig. 4.6.

> The diagonal triangle of a quadrangle inscribed in a conic is self-polar.

Figure 4.7 illustrates this situation: the diagonal triangle is indicated by the grey points and the tangents from two of its vertices are the grey lines.

The proof is simple if we choose the reference points and unit point to be the four vertices of the inscribed quadrangle. The equation of the conic then has the canonical form

$$dYZ + eZX + fXY = d + e + f = 0,$$

the vertices of the diagonal triangle are (011), (101) and (110) and its edges are

$$\begin{vmatrix} 1 & 0 & 1 \\ 1 & 1 & 0 \end{vmatrix} = [-1\ 1\ 1] \quad \text{etc.}$$

4.5 Self-Polar Triangles

Fig. 4.7 A quadrangle inscribed in a conic

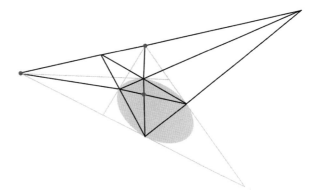

Fig. 4.8 A quadrilateral circumscribing a conic

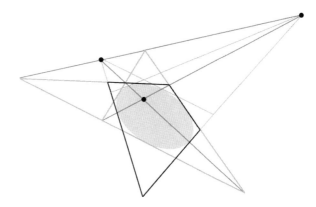

Using the prescription $L = P^T Q$ for the polar of a point P, the polars of the three vertices are given by the rows of

$$\begin{pmatrix} 0 & 1 & 1 \\ 1 & 0 & 1 \\ 1 & 1 & 0 \end{pmatrix} \begin{pmatrix} 0 & f & e \\ f & 0 & d \\ e & d & 0 \end{pmatrix} = \begin{bmatrix} -d & d & d \\ e & -e & e \\ f & f & -f \end{bmatrix}.$$

That is, they are the three lines $[-1\ 1\ 1]$, $[1\ -1\ 1]$ and $[1\ 1\ -1]$—the sides of the diagonal triangle.

The *dual* of this theorem is:

> *The diagonal triangle of a quadrilateral circumscribing a conic is self-polar.*

In Fig. 4.8 the circumscribing quadrilateral is indicated in black, the vertices of its diagonal triangle are marked.

These two theorems, and all the incidences that we have not considered, which appear as if by magic, seem at first bewilderingly complicated. But a little imagination can make it all seem 'self-evident'! Recall how projective geometry had its origins in the making of accurate drawings. These figures actually look like 'perspective drawings' of squares inscribed and circumscribed to a circle on an infinite

Fig. 4.9 Squares inscribed in and circumscribing a circle

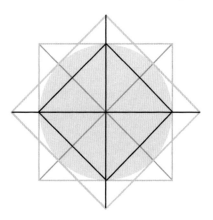

plane. The lines at the top of the figures are 'horizon lines' and the points on them are 'vanishing points' for families of parallel lines. Let us therefore choose this upper line as the *line at infinity* to get an *affine* version of the theorems. We then have parallelograms inscribed in and circumscribing an ellipse. Proceeding further to a Euclidean specialization we have squares inscribed in and circumscribing a circle. The two previous diagrams, superimposed, have become simply the Euclidean diagram Fig. 4.9.

4.6 Mutually Polar Triangles

Two triangles are mutually polar with respect to a conic if each edge of one is the polar of a vertex of the other.

> Two triangles are perspective if and only if they are mutually polar with respect to a conic.

Consider the polar triangle of the reference triangle, with reference to a conic with matrix

$$\begin{bmatrix} a & f & e \\ f & b & d \\ e & d & c \end{bmatrix}.$$

Its sides are given by the rows of this matrix. The points of intersection of corresponding sides of the two triangles are given by

$$\begin{vmatrix} 1 & 0 & 0 \\ a & f & e \end{vmatrix} = (0 \quad -e \quad f), \quad \begin{vmatrix} 0 & 1 & 0 \\ f & b & d \end{vmatrix} = (d \quad 0 \quad -f),$$

$$\begin{vmatrix} 0 & 0 & 1 \\ e & d & c \end{vmatrix} = (-d \quad e \quad 0),$$

which are collinear because

$$\begin{vmatrix} 0 & -e & f \\ d & 0 & -f \\ -d & e & 0 \end{vmatrix} = 0.$$

Conversely, a canonical form for a pair of perspective triangles is the reference triangle and a triangle whose vertices are given by the columns of a matrix of the form

$$\begin{pmatrix} \alpha & 1 & 1 \\ 1 & \beta & 1 \\ 1 & 1 & \gamma \end{pmatrix}.$$

It is easily verified that the two triangles are mutually polar with respect to the conic whose matrix Q is the inverse of this one.

Now consider the Desargues' (10_3) labeled as in Fig. 1.7. We have established that there is a conic Q with respect to which the points **14**, **24** and **34** are, respectively, the poles of the lines **235**, **135** and **125**. By repeated application of the result that 'if A lies on the polar of B then B lies on the polar of A', we deduce that **ij** and **klm** are pole and polar, where **ijklmn** is any permutation of **12345**:

> For any Desargues' (10_3) there is a conic with respect to which it is self-polar. Any pair of perspective triangles in it are mutually polar and the vertex and axis of the perspectivities are a pole-polar pair.

4.7 Metric Planes

This section is a rather interesting digression. Projective geometry is geometry without *metrical* properties. There is no such thing as the *distance* between two points and there is no meaning that can be given to the idea of the *angle* between two intersecting lines. These properties can be introduced by various specializations in much the same way that the projective plane can be specialized to the affine plane by selecting a line and calling it the 'line at infinity'.

Already in affine geometry there is a partial emergence of the length concept. As we saw in Sect. 3.4, the ratio of two segments of a line, or the ratio of two segments of two parallel lines, are well defined, though the comparison of length of line segments that are not parallel remains undefined. There is no angle measure.

A *metric plane* can be derived from a projective plane by singling out, not a line at infinity, but a *conic at infinity* (called the *ideal conic* or the *absolute conic* or simply the *absolute*), along with all its tangents, which are called *isotropic lines*. A consistent definition of distances and angles can then be made in terms of cross-ratios.

First, suppose that a real, non-singular conic Q is selected as the set of points 'at infinity'. As in affine geometry, lines are *parallel* if their point of intersection is at infinity—that is, on the absolute conic. Given any two non-infinite points A and B, the line AB will intersect Q at two points M and N, which may be real and distinct, real and coincident, or complex conjugates. The logarithm of the cross-ratio

$$(AB) = k \ln(AB, MN)$$

(k is an arbitrary factor that allows the unit of length to be chosen) satisfies

$$(AB) = -(BA)$$

and, for any three non-infinite collinear points A, B and C,

$$(AB) + (BC) = (CA).$$

Hence we can define the *length* of a line segment AB to be (AB), the *distance* between any two points A and B as the absolute value $|(AB)|$. Observe that isotropic lines are lines of zero length—the distance between any two points on an isotropic line is zero.

For similar reasons, for any two lines a and b there are two tangents m and n to Q through their point of intersection and the *angle* between a and b can be defined by

$$(ab) = (-i/2) \ln(ab, mn).$$

Observe that if the lines are parallel, i.e., intersecting on Q, then m and n coincide (a single isotropic line) and $(ab) = \ln(1) = 0$; the angle between parallel lines is zero. Two lines a and b are defined to be *orthogonal* if they are harmonic conjugates with respect to the pair of lines m, n. That is, for a pair of orthogonal lines

$$(ab) = (-i/2) \ln(-1) = (-i/2) \ln(e^{i\pi}) = \pi/2.$$

(The factor $-i\pi$ in the definition of (ab) was chosen so that this would correspond to the radian as the unit of angle. For angles measured in degrees it would be $-180i$.)

We get different kinds of metric plane, depending on the nature of the absolute conic Q and its associated conic envelope. To investigate in a completely general way, we can take the canonical form for the quadric to be $Q = 0$ where

$$Q = \varepsilon \eta X^2 - \varepsilon Y^2 + Z^2,$$

with the associated conic envelope $\tilde{Q} = 0$ where

$$\tilde{Q} = l^2 - \eta m^2 + \varepsilon \eta n^2.$$

For a non-singular Q, ε and η separately can take the values $+1$ or -1. The canonical equations have been written in this way so that we can also include the cases of metric planes with singular absolutes, by allowing ε or η, or both, to be zero. We then get nine different kinds of two-dimensional metric geometry as specializations of the projective geometry P(2, R), which we shall refer to as the metric planes $\{\varepsilon, \eta\}$. In each case the points of the metric plane are the real points for which $Q > 0$

4.7 Metric Planes

and the lines are the real lines for which $\tilde{Q} = 0$. We shall describe each of these geometries, very briefly.

The measure of length in $\{\varepsilon, \eta\}$ is referred to as *hyperbolic, parabolic* or *elliptic*, according as ε is positive, zero or negative, and the angular measure is referred to as *hyperbolic, parabolic* or *elliptic*, according as η is positive, zero or negative.

$\{--\}$: *Riemann's elliptic geometry.* The absolute is a non-singular virtual conic. All points and lines of P(2, R) except the points on Q and the isotropic lines belong to the geometry. Every line has a finite total length, the same for every line. There are no parallel lines. It is the geometry on the surface of a sphere with antipodal points identified—the great circles are the 'lines'.

$\{+-\}$: *Lobachevsky's hyperbolic geometry.* The absolute is a real non-singular conic. This is the first 'non-Euclidean' geometry to be discovered. The points of this metric geometry are the points of P(2, R) *inside* Q and the lines are all the lines that cut Q in a pair of real points. Angle measure is elliptic, like the angle measure in Euclidean geometry—the total angle around any point is finite. Two lines are parallel if they intersect on Q. There are an infinite number of lines parallel to a given line through a given point not on the line.

$\{-+\}$: *Anti-hyperbolic geometry.* The absolute is a real non-singular conic. But now the points are those points of P(2, R) *outside* Q and the lines are those lying entirely outside Q. A peculiar feature of this geometry is that there are pairs of points with no line belonging to the metric plane joining them.

$\{++\}$: *Doubly hyperbolic geometry.* The absolute is again real and non-singular. The points all lie outside it and the lines are those that cut Q in a pair of real points.

$\{0-\}$: *Euclidean geometry.* The canonical forms for the absolute are $Z^2 = 0 = l^2 + m^2 = 0$, so we have an *affine* geometry with extra structure: a pair of conjugate complex points (1 *i* 0) and (1 −*i* 0) on the line at infinity [001]. Length measure is parabolic—lengths of line segments can be consistently defined for each family of real parallel lines. We call the two special points (1 ±*i* 0) J and $\bar{\text{J}}$. A conic containing J and $\bar{\text{J}}$ has an equation of the form

$$X^2 + Y^2 + 2(dYZ + eZX) + cZ^2 = 0.$$

Since $Z \neq 0$ for all finite points, we can define *inhomogeneous* coordinates $(x, y) = (X/Z, Y/Z)$. The two axes are [010] and [100] which are orthogonal because they are the harmonic conjugates with respect to the isotropic lines [±*i* 1 0] through their intersection (001). In terms of these inhomogeneous *Cartesian* coordinates, a conic containing J and $\bar{\text{J}}$ is then

$$(x + d)^2 + (y + e)^2 = r^2,$$

where

$$r^2 = d^2 + e^2 - c.$$

The pole of the line at infinity with respect to this conic is $(-e, -d)$.

A *circle* is defined to be a conic containing the two 'circular points at infinity', J and J̄, and its *centre* is defined as the pole of the line at infinity. Comparison of lengths of *non-parallel* line segments can then be made if we take the radius of the unit circle $x^2 + y^2 = 1$ as a universal length unit.

The isotropic lines are the lines that pass through one or other of the points J and J̄, so there is a conjugate complex pair of isotropic lines through every point.

{0+}: *Pseudo-Euclidean geometry*. The canonical equations for the absolute are $Z^2 = l^2 - m^2 = 0$. We have a pair of *real* specialized points (1 ±1 0) on the line at infinity [001]. These special points partition the lines at infinity into two segments. The lines through any finite point are of two classes, separated by the two isotropic lines through the point. This metric plane is an analogue of the four-dimensional Minkowski space-time of special relativity. In this context the isotropic lines are called 'null lines' and the two kinds of non-null line are 'space-like' and 'time-like'.

{−0}: *Anti-Euclidean geometry*. The canonical equations for the absolute are $Y^2 + Z^2 = l^2 = 0$, which is therefore a conjugate complex pair of lines [0 1 ±i] and their (real) point of intersection (100).

{+0}: *Anti-pseudo-Euclidean geometry*. $-Y^2 + Z^2 = l^2 = 0$. A pair of *real* lines at infinity [0 1 ±1]. They partition the projective plane into two regions. The points of this metric plane are all the points of one of these regions. There is one isotropic line through every point (through the intersection of the lines at infinity).

{00}: *Galilean geometry*. $Z^2 = l^2 = 0$. This metric plane is an affine plane with a special point on the line at infinity. There is a therefore a single family of parallel isotropic lines. We set up two coordinate axes x and t, with t along an isotropic line. If the t-axis is regarded as a time-axis, then every line in the plane corresponds to a *uniform motion* along the x axis. Hence the name 'Galilean' geometry.

4.8 Pascal's Theorem

Pascal discovered a beautiful theorem about six points on a conic, when he was only 16 years old. He called the figure the *hexagrammum mysticum*:

> *If a hexagon is inscribed in a conic, the three intersections of pairs of opposite edges are collinear.*

There are various ways of proving this. Pascal's method of proof is not known. The simple proof here presupposes that the conic is not degenerate (if it is, the theorem is Pappus's theorem...) and that the six hexagon vertices on it are distinct. We can choose four of the vertices of the hexagon to be the vertices of the reference triangle and the unit point. The equation of the conic then reduces to

$$dYZ + eZX + fXY = 0, \qquad d + e + f = 0.$$

4.8 Pascal's Theorem

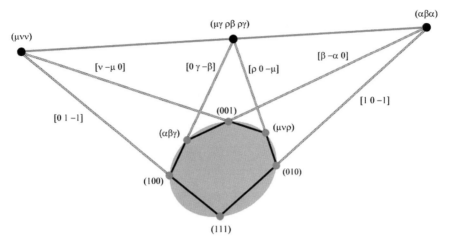

Fig. 4.10 Pascal's hexagon theorem

Since the conic is not degenerate, $def \neq 0$. Call the remaining two hexagon vertices $(\alpha\ \beta\ \gamma)$ and $(\mu\ \nu\ \rho)$.

The coordinates of the hexagon edges and of the points of intersections of the pairs of opposite edges are then easily obtained. They label Fig. 4.10. The determinant of the matrix whose rows are the coordinates of the three intersections is

$$\begin{vmatrix} \alpha & \beta & \alpha \\ \mu & \nu & \nu \\ \mu\gamma & \rho\beta & \rho\gamma \end{vmatrix}.$$

Observe that α cannot be zero because the condition for $(\alpha\ \beta\ \gamma)$ to lie on the conic is then $d\beta\gamma = 0$. But d is not zero so this implies β or γ zero. But then $(\alpha\ \beta\ \gamma)$ would be one of the three reference points, and we have supposed that the six points are distinct. Hence α cannot be zero. By the same argument, none of the coordinates $\alpha\beta\gamma\mu\nu\rho$ are zero.

Divide the first row of the determinant by α, the second by ν and the third by $\rho\gamma$. We get

$$\begin{vmatrix} 1 & \beta/\alpha & 1 \\ \mu/\nu & 1 & 1 \\ \mu/\rho & \beta/\gamma & 1 \end{vmatrix}.$$

Now divide the first column by μ and the second by β and we get

$$\begin{vmatrix} 1/\mu & 1/\alpha & 1 \\ 1/\nu & 1/\beta & 1 \\ 1/\rho & 1/\gamma & 1 \end{vmatrix}.$$

Fig. 4.11 Perspective triangles in relation to a conic

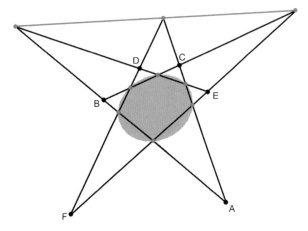

The determinant is zero because its three rows are linearly dependent,

$$d/\mu + e/\nu + f/\rho = 0$$
$$d/\alpha + e/\beta + f/\gamma = 0$$
$$d + e + f = 0,$$

so the three points are collinear.

(It is instructive to choose the 'Pascal line' in the diagram that illustrates Pascal's theorem as the 'line at infinity', to get an affine version, and then to draw a Euclidean version. The Euclidean theorem is 'if two pairs of opposite sides of a hexagon inscribed in a circle are parallel, so is the remaining pair'.)

Figure 4.11 reveals the diagram illustrating Pascal's theorem in a different light. There is a pair of *perspective triangles* ABC and DEF!

The presence of these triangles enables the *converse* of Pascal's theorem to be stated in the form:

> *The six intersections of pairs of* non-*corresponding sides of a pair of perspective triangles lie on a conic.*

Specifically, if ABC and DEF are two triangles in perspective then the six points BC·FD, BC·DE, CA·DE, CA·EF AB·EF and AB·FD lie on a conic. To prove this, let the vertex of the perspectivity be the unit point and choose ABC to be the reference triangle. Then the coordinates of the sides of the other triangle are given by the rows of a matrix of the form

$$\begin{bmatrix} \alpha & 1 & 1 \\ 1 & \beta & 1 \\ 1 & 1 & \gamma \end{bmatrix}.$$

The six points of intersection of non-corresponding sides are then

$$(0\ 1\ -\beta)(0\ -\gamma\ 1)(-\gamma\ 0\ 1)(1\ 0\ -\alpha)(1\ -\alpha\ 0)(-\beta\ 1\ 0)$$

Fig. 4.12 Brianchon's theorem

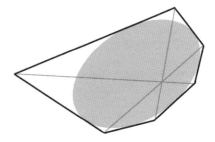

which, as is easily verified, all lie on the conic whose matrix Q is

$$\begin{pmatrix} 2\alpha & 1+\alpha\beta & 1+\gamma\alpha \\ 1+\alpha\beta & 2\beta & 1+\beta\gamma \\ 1+\gamma\alpha & 1+\beta & 2\gamma \end{pmatrix}.$$

The dual of Pascal's theorem is Brianchon's theorem (Fig. 4.12):

The three diagonals of a hexagon circumscribing a conic are concurrent.

4.9 The Extended Pascal Figure

There are sixty different hexagons with the same six vertices (which we shall call **1**, **2**, **3**, **4**, **5** and **6**), according to the order in which the vertices are traversed. Cyclic permutations of **123456** and reversal of the sequence describe the same hexagon; hence there are $6!/(6 \times 2) = 60$ hexagons. Hence six points on a conic give rise to *sixty Pascal lines* each containing three points of intersection of pairs of opposite sides, for some hexagon.

The sides of all these hexagons are the lines joining pairs of the six points. There are $\binom{6}{2} = 15$ of them. Now, 15 general lines meet in pairs at $\binom{15}{2} = 105$ points, but in this particular case there are six points where five of the lines meet—the six points on the conic. Hence, apart from these six there are $105 - 6 \times \binom{5}{2} = 45$ points of intersection of the 15 lines. We have, then, a $(45_2 15_6)$ of points and lines. In Fig. 4.13 they are the black points and the black lines (unfortunately, five of the points are off the page...). A few of the Pascal lines have been indicated by paler lines.

Each Pascal line, of course, contains three of these points, so the black points and the Pascal lines constitute a $(45_4 60_3)$. Just one of the points of concurrence of four Pascal lines is indicated in the figure.

All this comes about in a straightforward way from Pascal's theorem, as soon as it is realized that many hexagons can have the same six vertices. It should not be too surprising. The surprise is that there are *other* points where *three* Pascal lines are concurrent:

The sixty Pascal lines are concurrent in threes through twenty points.

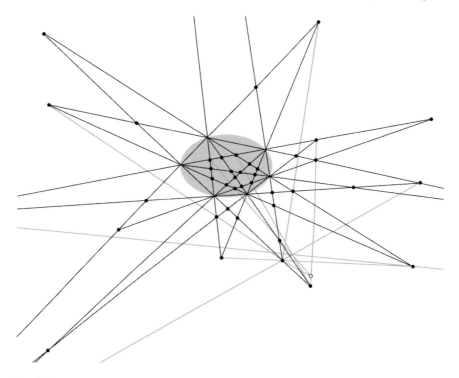

Fig. 4.13 Pascal lines

Fig. 4.14 One of the sixty hexagons determined by six points on a conic

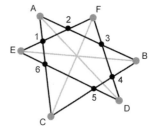

One of them has been indicated by the small circle in Fig. 4.13. They are called the *Steiner points*. To prove this, consider Fig. 4.14, which is a schematic representation of any one of the sixty hexagons, labeled **123456**. The two triangles ABC and DEF are perspective from the Pascal line of this hexagon. They are therefore perspective from a point—the three lines AD, BE and CF in Fig. 4.14 are concurrent. But these three lines are the Pascal lines for the hexagons **143256**, **123654** and **163452** (because AD contains the intersections A = **32 · 61** and D = **43 · 56** of pairs of opposite sides of **143256**, and so on...). Observe also that the three hexagons **143256**, **123654** and **163452** are obtained from **123456** by keeping one set of alternate vertices fixed, reversing the order of the other set of alternate vertices and cyclically permuting it.

4.9 The Extended Pascal Figure

Starting from just six points on a conic, it is astonishing that a formidable structure of points and lines, with surprising incidence properties, can be constructed. The discoveries were made by various eminent mathematicians, whose names are attached to the various kinds of point and line. We shall simply describe the structure here, without further attempts at proofs. All the statements *could* in principle be verified by using the elementary methods of homogeneous coordinates to find the line through a point pair, or the intersection of a line pair—but that would be laborious. The S_6 symmetry ensures that only one example of each kind of incidence needs be proved—the rest follow from permutations of **123456**. In case anyone wishes to try it, a fruitful approach is to use the canonical form $Y^2 = XZ$ for the conic and (without loss of generality) to let the six points **123456** be

$$(1\ 0\ 0)(0\ 0\ 1)(1\ 1\ 1)(\mu^2\ \mu\ 1)(\nu^2\ \nu\ 1)(\rho^2\ \rho\ 1).$$

By permuting $\mu\nu\rho$ and **456** all 15 edges are immediately deducible from these seven:

$$23 = [1\ -1\ 0] \qquad 31 = [0\ 1\ -1] \qquad 12 = [1\ 1\ 0]$$
$$14 = [0\ -1\ \mu] \qquad 24 = [-1\ \mu\ 0] \qquad 34 = [-1\ \mu+1\ -\mu]$$
$$56 = [-1\ \nu + \rho\ \nu\rho].$$

Sylvester's duads and synthemes provide an elegant way of naming the points and lines of the extended *hexagrammum mysticum*. The edges of **142536** are of course the six lines **14 42 25 53 36 61**. These six duads belong to the two *synthemes* **14.25.36** and **42.53.61**. Consulting the table of synthemes, we see that these are *24* and *25*. In this way, any one of the 60 hexagons can be specified by a pair of duads with a common digit (which refers us to two synthemes in the same row of the table—which have no duad in common and therefore together specify six hexagon edges). In this way, the symbol *24.25* provides a concise way of specifying the hexagon **142536** (provided, of course, that we have the table of synthemes ready to hand for doing the translation). In this scheme, *32.34*, for example, refers us to the two synthemes **34.56.12** and **45.16.23**, and hence the hexagon **123456**. The same notation can be used to label the Pascal lines associated with the hexagons: the *Pascal line 32.34* is the line containing the three points **12** · **45**, **23** · **56** and **45** · **61**.

Now look again at the three Pascal lines concurrent at a Steiner point suggested by Fig. 4.14. They are the Pascal lines of the three hexagons **143256**, **123654** and **163452**. Employing the syntheme table in the manner just described, they are *65.61*, *56.51* and *16.15*. Observe the pattern. The Steiner point shared by these three Pascal lines can be unambiguously denoted by the triple *165*. We get a neat naming of the Steiner points by the $\binom{6}{3} = 20$ triples. Example: the three Pascal lines that are concurrent at the Steiner point *123* are *12.13*, *23.21* and *31.32*.

The Steiner points lie in fours on 15 Plücker lines, *forming a* $(20_3 15_4)$.

(See Fig. 3.5.) For example *123, 124, 134* and *234* lie on a Plücker line which I shall denote as *1234*.

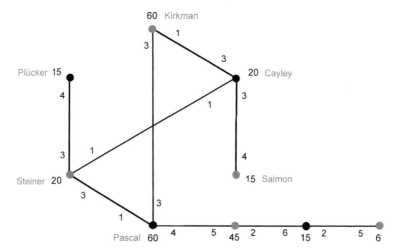

Fig. 4.15 Incidences in the *hexagrammum mysticum*

There are other 'surprise' concurrences of three Pascal points—there are sixty *Kirkman points*:

> The Kirkman points and Pascal lines form a configuration (60_3), which consists of three separate Desargues' (10_3) configurations.

Just one Kirkman point is indicated (by the white circle) in Fig. 4.13. Three Pascal lines concurrent at a Kirkman point have the generic form **42.43 43.41 41.42**. I shall call this particular Kirkman point *45.46* so that the notation for the sixty lines and sixty planes will match. (Translation between the two nomenclatures is straightforward—it uses the fact that **123** and *456* are complementary triads. Its purpose is only to simplify the notation and to emphasize the fact that there is a mysterious kind of reciprocity in the extended Pascal configuration.)

> The Kirkman points lie in threes on twenty Cayley lines.

For example, *12.13*, *23.21* and *31.32* are collinear on a Cayley line which may be denoted by *123*.

> Every Cayley line contains one Steiner point.

For example, **123** and *456* are incident.

> The Cayley lines are concurrent in fours through 15 Salmon points, forming a $(15_4 20_3)$.

(See Fig. 3.12.) For example, *123*, *124*, *134* and *234* are concurrent at a Salmon point which may be denoted by *1234*.

Figure 4.15 is a summary of all this—a kind of Levi graph in which each grey vertex represents many points and each black vertex represents many lines.

4.10 Quadrics

A *quadric surface* (or, simply, a *quadric*) in a three-dimensional projective space $P(3, R)$ is the set of all points P whose four homogeneous coordinates $P = (x^1 \ x^2 \ x^3 \ x^4)$ (written as a column) satisfy an equation of the form

$$P^T Q P = 0$$

where Q is a real 4×4 symmetric matrix.

By diagonalizing Q, as we did in the two-dimensional case of a non-singular conic, we find three canonical forms for a real non-singular quadric:

$$(x^1)^2 + (x^2)^2 + (x^3)^2 + (x^4)^2 = 0$$
$$(x^1)^2 + (x^2)^2 + (x^3)^2 - (x^4)^2 = 0$$
$$(x^1)^2 + (x^2)^2 - (x^3)^2 - (x^4)^2 = 0.$$

The first of these forms is a *virtual* quadric, which contains only complex points. The third form is obtained from a *ruled quadric*. To see why it is so named, write $y^1 = x^4 + x^1$, $y^2 = x^2 + x^3$, $y^3 = x^2 - x^3$, $y^4 = x^4 - x^1$ and the equation $(x^1)^2 + (x^2)^2 - (x^3)^2 - (x^4)^2 = 0$ becomes

$$y^1 y^4 = y^2 y^3.$$

With this canonical form every point on the quadric has the form

$$(\theta\phi \ \ \theta \ \ \phi \ \ 1).$$

If we select any point on the quadric, we can keep θ fixed and let ϕ vary. We get a *real line* through the chosen point, which *lies entirely in the quadric*. Or we can keep ϕ fixed and vary θ and get another real line lying entirely in the conic. These lines are *generators* of the quadric. There are two through every point $(\theta_0\phi_0 \ \theta_0 \ \phi_0 \ 1)$ of the quadric, one consisting of all the points $(\theta\phi_0 \ \theta \ \phi_0 \ 1)$ and one consisting of all the points $(\theta_0\phi \ \theta_0 \ \phi \ 1)$. (See Fig. 4.16. In the *affine*—or Euclidean—classification these are a 'hyperbolic paraboloid' and an 'elliptic hyperboloid of one sheet'. Projective geometry recognizes no such distinction.)

If complex coordinates are permitted, *any* non-singular quadric can be transformed to the canonical form $y^1 y^4 = y^2 y^3$ and so:

> On a non-singular quadric there are two lines through every point of the surface which lie entirely in the surface.

For a non-ruled quadric, these *generators* are complex. For a ruled quadric they are real. A *singular* quadric has one of the canonical forms

$$(x^1)^2 + (x^2)^2 + (x^3)^2 = 0$$

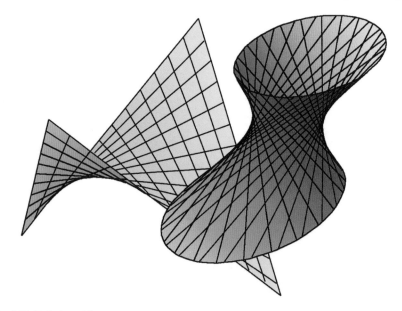

Fig. 4.16 Ruled quadrics

$$(x^1)^2 + (x^2)^2 - (x^3)^2 = 0$$
$$(x^1)^2 + (x^2)^2 = 0$$
$$(x^1)^2 - (x^2)^2 = 0$$
$$(x^1)^2 = 0,$$

corresponding, respectively, to: a virtual *quadric cone*, a real quadric cone, a conjugate complex pair of real planes, a pair of real planes, and a single plane.

4.11 Pascal's Theorem Again

A neat proof of Pascal's theorem follows from the existence of the two reguli on a ruled quadric. A hexagon can be inscribed on a ruled quadric so that all its edges are generators of the quadric—like the hexagon **123456** in Fig. 4.17. Let L be the line of intersection of the planes **123** and **456**; let M be the line of intersection of the planes **345** and **612** and let N be the line of intersection of the planes **561** and **234**. These three lines are concurrent in pairs: M · N = **61** · **34**, N · L = **23** · **56** and L · M = **12** · **45**. L, M and N therefore lie in a plane P. The section of the quadric by any other plane P′ is a conic, and the intersections of P′ with the six generators **12 23 34 45 56 61** are *six vertices of a hexagon inscribed in the conic*. The sides of this planar hexagon are the intersections of P′ with the planes **123, 234, 345, 456,**

4.12 Tangent planes

Fig. 4.17 A hexagon inscribed on a quadric

561 and **612**. The intersections of pairs of opposite sides of the planar hexagon are therefore the intersections of P′ with L, M and N, which all lie on the line P · P′.

4.12 Tangent planes

The intersection of a plane and a quadric is a conic. The plane is a *tangent plane* if this conic is degenerate, i.e., a pair of generators. The intersection of these two generators is the *point of contact* of the tangent plane and the quadric.

The set of all tangent planes to a non-singular quadric with matrix Q is given by

$$PQ^{-1}P^T = 0,$$

where P is the set of homogeneous coordinates for the planes (written as a row).

Let P be a plane satisfying this equation. Then $A = Q^{-1}P^T$ lies on the quadric and also on the plane. Let B be another point on the intersection of the plane and the quadric. The condition for $A + \lambda B$ to lie on the intersection is

$$0 = (A + \lambda B)^T Q (A + \lambda B) = 2\lambda A^T Q B.$$

But this is true for all λ because $A^T Q B = PB$. The conic of intersection of the plane and the quadric contains the line consisting of all points $A + \lambda B$. So it is a degenerate conic—any plane satisfying $PQ^{-1}P^T = 0$ is a tangent to the quadric Q. Conversely, if P is a tangent to Q there are two lines $A + \lambda B$ and $A + \mu C$ in P that also lie on Q. It follows that $A^T Q A = A^T Q B = A^T Q C = 0$. Therefore the plane $A^T Q$ contains A, B and C. But A, B and C define the plane uniquely. Hence $P = A^T Q$, which satisfies $PQ^{-1}P^T = 0$.

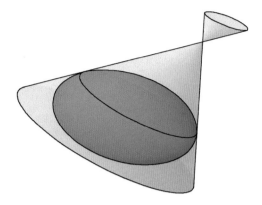

Fig. 4.18 A quadric cone generated by tangent lines to a quadric

4.13 Polarity

By analogy with the two-dimensional case, we define the *polar plane* (or, simply, the polar) of a point A with respect to a non-singular quadric Q to be $P = A^T Q$, and we define the *pole* of a plane P with respect to Q to be the point $A = Q^{-1} P^T$. Then the polar plane of a point A *on* Q is the tangent plane at A (because it satisfies $P Q^{-1} P^T = 0$).

In the two-dimensional case, we saw that the polar of a point A with respect to a conic is the line joining the two points of contact of the two tangents through A. The three-dimensional case is not so simple—there are infinitely many tangent planes (and tangent lines) to a quadric through a given point. However, consider all the *lines* through A that are tangents to a quadric Q. As for a conic, a line and a quadric have two points in common, which may be real, coincident (a single real point), or conjugate complex, and a line is *tangential* to a quadric Q if it touches it at a single point. For the line $A + \lambda B$ to be tangential to Q, the condition is (in complete analogy with the two-dimensional case)

$$\left(B^T Q A\right)^2 - \left(B^T Q B\right)\left(A^T Q A\right) = 0.$$

Taking B to be the point of contact of the tangent line with the quadric we have $B^T Q B = 0$ and hence $B^T Q A = A^T Q B = PB = 0$: B lies on the polar plane, $P = A^T Q$, of A.

> *The points of contact of the tangent lines to a quadric from a point* A *all lie in the polar plane of* A.

(Figure 4.18.) These contact points of course all belong to the conic in which the polar plane of A cuts the quadric. All the tangent lines through A generate a quadric cone, with vertex at A.

4.14 Affine Classification of Quadrics

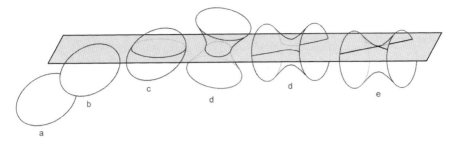

Fig. 4.19 Intersections of quadrics by a plane

4.14 Affine Classification of Quadrics

An affine 3-space is obtained from P(3, R) by selecting a plane at infinity. As shown in Fig. 4.19, there are five kinds of non-degenerate quadric in affine 3-space, depending on the nature of the intersection of the quadric with the plane at infinity (the intersection is, of course, a conic).

The equation of a *non-ruled* quadric may be given the canonical form

$$(x^1)^2 + (x^2)^2 + (x^3)^2 - (x^4)^2 = 0$$

and there are three possibilities for its conic at infinity.

(a) *Ellipsoid.* The conic at infinity is a virtual conic—the quadric has no real infinite points. The plane at infinity can be taken to be [0001] (i.e., $x^4 = 0$). Inhomogeneous coordinates for the affine space can be defined by (X Y Z) = $(x^1/x^4 \; x^2/x^4 \; x^3/x^4)$. The quadric has the canonical form

$$X^2 + Y^2 + Z^2 = 1.$$

(b) *Elliptic paraboloid.* The conic at infinity is a pair of conjugate complex lines; the plane at infinity is tangential to the quadric at their (real) point of intersection. Take the plane at infinity to be [0011] ($x^3 + x^4 = 0$). Homogeneous coordinates for this plane can be chosen to be $(y^1 \; y^2 \; y^3) = (x^1 \; x^2 \; x^3 - x^4)$. The equation of the conic is $(y^1)^2 + (y^2)^2 = (y^1 + iy^2)(y^1 - iy^2) = 0$. It is a pair of conjugate complex lines [1 $\pm i$ 0], intersecting in the real point $(y^1 \; y^2 \; y^3) = (001)$, or $(x^1 \; x^2 \; x^3 \; x^4) = (0 \; 0 \; 1 \; -1)$. The quadric is tangential to the plane at infinity, touching it at a single real point. Inhomogeneous coordinates for the affine space are (X Y Z) = $(x^1/(x^3 + x^4) \; x^2/(x^3 + x^4) \; (x^3 - x^4)/(x^3 + x^4))$, and the canonical equation of the quadric is

$$X^2 + Y^2 + Z = 0.$$

(c) *Elliptic hyperboloid with two sheets.* The conic at infinity is a real non-degenerate conic. Let the plane at infinity be [0010] ($x^3 = 0$). The quadric and the plane at infinity intersect in a real, non-degenerate conic $(x^1)^2 +$

Fig. 4.20 Two kinds of hyperboloid, sharing the same asymptotic cone

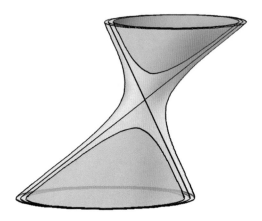

$(x^2)^2 - (x^4)^2 = 0$. Inhomogeneous coordinates for the affine space: $(X\ Y\ Z) = (x^1/x^3\ x^2/x^3\ x^4/x^3)$, and the equation of the quadric is

$$X^2 + Y^2 - Z^2 + 1 = 0.$$

(NB: to avoid any possibility of confusion, observe that I am using (X Y Z) for *inhomogeneous* coordinates in affine 3-space, and *italic* *(X Y Z)* for *homogeneous* coordinates in a projective plane.)

For a *ruled* quadric, the canonical equation in P(3, R) is

$$(x^1)^2 + (x^2)^2 - (x^3)^2 - (x^4)^2 = 0.$$

In affine space, there are two possibilities

(d) *Elliptic hyperboloid with one sheet.* The conic at infinity is real and non-degenerate. The plane at infinity can be taken to be [0001] ($x^4 = 0$). (In Fig. 4.19 I have illustrated this in two different ways. They are fully equivalent, since the plane at infinity is a projective plane...) In terms of the inhomogeneous coordinates $(X\ Y\ Z) = (x^1/x^4\ x^2/x^4\ x^3/x^4)$ of the affine space, the equation of the quadric is

$$X^2 + Y^2 - Z^2 = 1.$$

(e) *Hyperbolic paraboloid.* The conic at infinity is a pair of real lines. Let the plane at infinity be [0110] ($x^2 + x^3 = 0$). Homogeneous coordinates for this plane can be chosen to be $(y^1\ y^2\ y^3) = (x^1\ x^4\ x^2 - x^3)$. The equation of the conic is $(y^1)^2 - (y^2)^2 = (y^1 + y^2)(y^1 - y^2) = 0$. It is a pair of real lines. The plane at infinity is a tangent plane, intersecting it in a pair of generators.

Taking $(X\ Y\ Z) = (x^1/(x^2 + x^3)\ x^4/(x^2 + x^3)\ (x^2 - x^3)/(x^2 + x^3))$ as inhomogeneous coordinates for the affine plane, the equation of the quadric is

$$X^2 - Y^2 + Z = 0.$$

The *centre* of an affine quadric is the pole of the plane at infinity. The lines joining the centre to the conic at infinity (which are tangential to the quadric, touching it

4.15 Reguli

at infinity) generate a cone, with vertex at the centre. This is the *asymptotic cone*. The centre is a real, non-infinite point only for the ellipsoid and the two kinds of hyperboloid, and the asymptotic cone is real only for the two kinds of hyperboloid (Fig. 4.20).

4.15 Reguli

Before proceeding, a little more needs to be said about the system of lines on a non-singular quadric.

Two or more lines in projective 3-space are said to be *skew* if they have no intersections (equivalently, if no two of them are coplanar). A *transversal* to a set of skew lines is a line that intersects all of them.

> *There is a unique transversal to any pair of skew lines, through any point that does not lie on either of them.*

This is fairly obvious: the unique transversal to two skew lines M and N through a point A is the line AB, where B is the intersection of N with the plane AM.

It follows that, if we are given three skew lines, there is an infinite family of transversals to all three of them, with one member of the family passing through each point of each of the three skew lines. A family of transversals to three skew lines is called a *regulus*. All members of a regulus are skew to each other because, if any two were coplanar this would imply coplanarity of the original three lines. The original three skew lines are transversals to all the lines of this regulus, so they themselves belong to a regulus. We thus have *a complementary pair of reguli*, each line of one being transversal to all the lines of the other. Thus we see that three skew lines uniquely determine a complementary pair of reguli. From this it follows that

> *There is a unique quadric through three skew lines.*

Proof Figure 4.21 represents three skew lines L, M and N and three of their transversals. We can choose the four reference points to be the four non-coplanar intersections, as indicated. By adjusting arbitrary overall factors associated with homogeneous coordinates we can assign the coordinates (1 1 0 0) and (1 0 1 0) to the two points so labeled, without any loss of generality. A *general* point on M is (0 0 1 μ) and a general point on the transversal through (0 1 0 0) and (0 0 0 1) is (0 1 0 ν). The point A is on the line through (1 1 0 0) and (0 0 1 μ) and therefore has the form (1 1 α $\alpha\mu$). It is also on the line through (1 0 1 0) and (0 1 0 ν) and so A has the form (1 β 1 $\beta\nu$). Hence $\alpha = \beta = 1$, $\mu = \nu$, and A is (1 1 1 μ). The unique matrix Q satisfying $P^T Q P = 0$, where P is any of these nine points, is

$$Q = \begin{pmatrix} 0 & 0 & 0 & 1 \\ 0 & 0 & -\mu & 0 \\ 0 & -\mu & 0 & 0 \\ 1 & 0 & 0 & 0 \end{pmatrix}.$$

Fig. 4.21 Three skew lines and three transversals

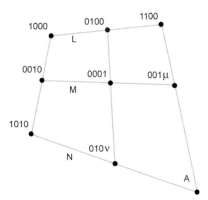

Fig. 4.22 Gallucci's theorem

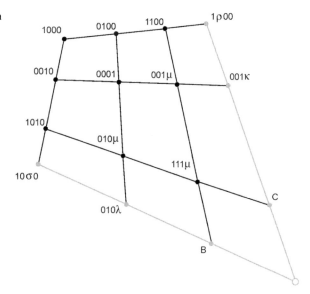

This is the matrix of the quadric containing the three skew lines.

A related theorem is:

> *If three skew lines all meet three other skew lines, any transversal to the first set of three meets any transversal to the second set of three.*

This result is known as *Gallucci's theorem* or the *16 point theorem*. Once it is recognized that the eight lines are generators of a quadric, this result begins to seem obvious. An elementary proof that doesn't refer to a quadric, goes like this:

The two sets of three skew lines are represented by the black lines in Fig. 4.22. The homogeneous coordinates labeling their nine points of intersection have already been established. The grey lines are the two general transversals, with homogeneous coordinates labeling them. The point B lies on the line through (1 0 σ 0) and (0 1 0 λ), so has the form (1 α σ $\alpha\lambda$). It is also on the line joining (1 1 0 0) and (0 0 1 μ) so has the form (1 1 β $\beta\mu$). Hence $\alpha = 1$, $\beta = \sigma$ and $\lambda = \sigma\mu$. B is then (1 1 σ $\sigma\mu$). Similarly, C is (1 ρ 1 $\rho\mu$) and $\kappa = \rho\mu$. A general point on the line through (1 ρ 0 0) and (0 0 1 κ) is (1 ρ γ $\gamma\kappa$) = (1 ρ γ $\gamma\rho\mu$) and a general point on the line through (1 0 σ 0) and (0 1 0 λ) is (1 δ σ $\delta\lambda$) = (1 δ σ $\delta\sigma\mu$). The two lines therefore intersect at (1 ρ σ $\rho\sigma\mu$).

4.16 Metric Spaces

Specializing P(3, R) to obtain three-dimensional metric spaces is analogous to the two-dimensional case, but with some interesting new features. A quadric is singled out as the *quadric at infinity*, or the *absolute*. In projective 3-space the canonical form for the equations of a non-singular quadric and its envelope (which is its collection of tangent planes) can be taken to be $Q = 0$ and $\tilde{Q} = 0$, where

$$Q = -\varepsilon\eta\xi(x^1)^2 + \varepsilon\eta(x^2)^2 - \varepsilon(x^3)^2 + (x^4)^2 = 0$$
$$\tilde{Q} = (p_1)^2 - \xi(p_2)^2 + \eta\xi(p_3)^2 - \varepsilon\eta\xi(p_4)^2 = 0$$

with ε, η and ξ taking the values ± 1. The prescription is extended to non-singular absolutes by also allowing ε, η or ξ to be zero. We then have *twenty-seven* different possibilities for three-dimensional metric spaces. The points of the metric space $\{\varepsilon\ \eta\ \xi\}$ are those for which $Q > 0$ and the planes are those for which $\tilde{Q} > 0$. The two equations for the absolute determine the points at infinity, the isotropic lines and the isotropic planes. These establish, through cross-ratios, lengths, angles between pairs of lines and angles between pairs of planes. It then turns out that the measure of distance between two points is elliptic, parabolic or hyperbolic according to whether ε is -1, 0 or 1; η has the same role for the measure of angles between intersecting lines, and ξ for angles between pairs of planes.

A description of each of the 27 cases one-by-one would become tedious. The principle is what is important. Considering that projective geometry is what emerges from the decision to abandon all metric properties, it is truly remarkable that a projective space contains within itself many kinds of metric geometry. The key to it all is the projective invariance of cross-ratios.

The familiar 3-dimensional Euclidean geometry is $\{0 - -\}$. The absolute for all the geometries $\{0\ \eta\ \xi\}$ is a *plane with a conic in it*, so in each of these cases there is a plane at infinity that is *itself* a metric plane; specifically, it is a $\{\eta\ \xi\}$. In particular, for three-dimensional Euclidean space, 'infinity' has the metrical properties of Riemann's elliptic geometry $\{- -\}$, which, as noted earlier, is that of a spherical surface with antipodal points identified. The angle between two lines through a finite point in Euclidean 3-space is a measure of the 'distance' between the two points at infinity on those lines. The angle between two planes through a line in Euclidean 3-space is the angle between the two lines at infinity on those planes (which can be imagined as two great circles on a 'sphere' at infinity). A *sphere* in Euclidean space is a real quadric whose intersection with the plane at infinity is the (virtual) conic at infinity, and the *centre* of a sphere is the pole of the plane at infinity.

For the eight metric geometries for which none of the parameters ε, η or ξ is zero, the absolute Q is a non-singular quadric, so that through every point at infinity there are two lines at infinity—generators of Q. These are a conjugate complex pair of lines or a real pair of lines according to the nature of Q. The existence of a complementary pair of reguli at infinity leads to some fascinating geometrical ideas, which are the subject of the next section.

4.17 Clifford Parallels

The discussion in this section could apply to any metric geometry with a non-singular absolute, but in some cases the generators of Q are *virtual* (i.e., complex rather than real) lines. For clarity and simplicity, let us suppose all the generators of Q—the lines at infinity—are real. Those metric spaces with a *real ruled quadric* as absolute are $\{+++\}$, $\{+-+\}$ and $\{-+-\}$. They can be taken together as a single geometry if we dispense with the restrictions $Q > 0$ and $\tilde{Q} > 0$; we can look into the geometry that consists of *all* the points of $P(3, R)$ that are not on the absolute, and *all* the non-isotropic planes. This metric space is *Clifford space*. In this space a special kind of parallelism for lines can be defined with the property that through any point there are *just two* lines 'parallel' to a given line (not containing the point).

The generators of Q are 'lines at infinity'. Two of them pass through every point at infinity. The two complementary reguli on Q may be called the *left* and the *right* regulus, consisting respectively of *left* generators and *right* generators.

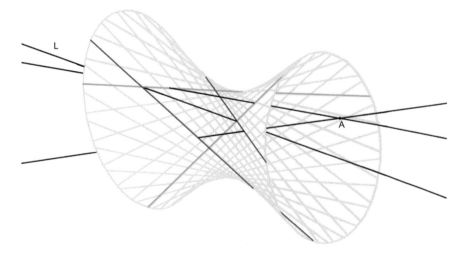

Fig. 4.23 The quadric at infinity in Clifford space

4.17 Clifford Parallels

Fig. 4.24 A construction in Clifford space based on a pair of intersecting lines

Any non-isotropic line of the geometry cuts Q at two real points, through each of which pass a right generator and a left generator. In other words, it intersects a pair of left generators and a pair of right generators.

Two lines are defined to be *left parallel* if both of them are transversals to the same pair of left generators, and are *right parallel* if both of them are transversals to the same pair of right generators. Then:

For any given line there is, through a given point, just one line that is left parallel to it and one line that is right parallel to it.

Figure 4.23 shows the left parallel and the right parallel to the line L through the point A.

The set of all lines transversal to the same pair of left generators is a *system of left parallel lines*. They are, of course, all left parallel to each other, and there is a unique line belonging to the family through every point of the space.

The set of all lines transversal to the same pair of right generators is a *system of right parallel lines*. They are all right parallel to each other, and there is a unique line belonging to the family through every point of the space.

Any pair of intersecting non-isotropic lines lies in a unique pair of quadrics whose reguli are a family of lines all left parallel to each other, and a family of lines all right parallel to each other.

This is an immediate consequence of Gallucci's theorem. Figure 4.24 is equivalent to Fig. 4.23, but now with a particular interpretation. A pair of intersecting lines is shown black. The four light grey lines are 'at infinity'—they are generators of the absolute. The black and light grey lines in the figure satisfy the conditions for Gallucci's theorem: three skew lines all meeting three other skew lines. Hence any pair of transversals, like the two darker grey lines in the figure, intersect. All these transversals therefore belong to a complementary pair of reguli, and hence all lie in a quadric. (Incidentally, note that this quadric and the absolute quadric have four lines in common. In general, two quadrics intersect in a quartic curve—in this case the quartic curve is degenerate, consisting of four lines. All the lines of one regulus

Fig. 4.25 As Fig. 4.24, with left and right parallelism interchanged

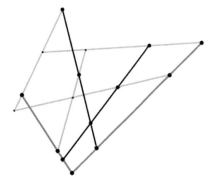

intersect the same pair of left generators of the absolute and hence are all left parallel to each other, and similarly all the lines of the other regulus are right parallel to each other. There are *two* situations, and hence two different quadrics, because the roles of the two black lines may be interchanged. The alternative situation is represented by Fig. 4.25.)

4.18 Isometries

The *isometries* in a metric geometry are the transformation of the metric space that preserve lengths and angles. Two figures are *congruent* if there is an isometry that maps one to the other. In the context of projective 3-space and the derivation of metric spaces from it, the isometries of a metric space are the *collineations that keep its absolute quadric invariant*. In Clifford space (which includes within itself the geometries $\{+++\}$, $\{+-+\}$ and $\{-+-\}$) the absolute can be taken in the canonical form

$$(x^1)^2 + (x^2)^2 - (x^3)^2 - (x^4)^2 = 0,$$

and so every collineation with (real) matrix S satisfying

$$SQS^T = Q, \quad Q = \begin{pmatrix} 1 & & & \\ & 1 & & \\ & & -1 & \\ & & & -1 \end{pmatrix}$$

is an isometry. Because of the irrelevance of an overall factor in homogeneous coordinates the matrices S may be taken to satisfy $|S| = \pm 1$. These S constitute the pseudo-orthogonal group $O(2, 2, R)$. If $|S| = 1$ we have a *direct* isometry and if $|S| = -1$ we have an *opposite* isometry.

Clifford space has the special distinction that the coordinates $(y^1 \; y^2 \; y^3 \; y^4)$ appear in the alternative canonical form. The absolute in this case can be taken in the canonical form

$$y^1 y^4 = y^2 y^3$$

4.18 Isometries

are *real*. The way in which the isometries express themselves in terms of these coordinates is of special interest. This canonical form is simply

$$|Y| = 0, \quad Y = \begin{pmatrix} y^1 & y^2 \\ y^3 & y^4 \end{pmatrix}.$$

The absolute is then invariant under the transformations

$$Y \to LYR$$

where L and R are real, non-singular 2 × 2 matrices, which may be taken to be *unimodular* because an overall factor in any set of homogeneous coordinates $(y^1\ y^2\ y^3\ y^4)$ has no significance. (We have here the group isomorphism $O(2,2,R) \sim SL(2,R) \otimes SL(2,R)$.)

Now write

$$L = \begin{pmatrix} \alpha & \beta \\ \gamma & \delta \end{pmatrix}, \quad \alpha\beta - \gamma\delta = 1;$$

$$R = \begin{pmatrix} \alpha' & \gamma' \\ \beta' & \delta' \end{pmatrix}, \quad \alpha'\beta' - \gamma'\delta' = 1.$$

When acting on the absolute, which in parametric form is now

$$Y = \begin{pmatrix} \theta\phi & \theta \\ \phi & 1 \end{pmatrix},$$

the action of L is a *homography* on the generators ϕ = constant,

$$\theta \to \frac{\alpha\theta + \beta}{\gamma\theta + \delta}, \quad \phi \to \phi$$

and the action of R is a homography on the generators θ = constant,

$$\phi \to \frac{\alpha'\phi + \beta'}{\gamma'\phi + \delta'}, \quad \theta \to \theta.$$

L may be called a *left translation*, and R a *right translation*. Obviously, any left translation commutes with any right translation. Any product of a left and a right translation is a *direct isometry*. The absolute is also invariant under the interchange of the two parameters, $\phi \leftrightarrow \theta$. Hence $y^2 \leftrightarrow y^3$ (A $\leftrightarrow$ A^T) is also an isometry. A combination of a direct isometry and this interchange is an *opposite* isometry.

We have dealt at length with homographies on the generators of the absolute of Clifford space (which includes within itself the geometries $\{+++\}$, $\{+-+\}$ and $\{-+-\}$) because then we don't have to take into consideration reality conditions. The canonical coordinates $(y^1\ y^2\ y^3\ y^4)$ are, like $(x^1\ x^2\ x^3\ x^4)$, all real. This is not so for other metric geometries. Consider, for example *elliptic space* $\{---\}$. The coordinates of the points $(x^1\ x^2\ x^3\ x^4)$ are real numbers, but the 'points at infinity' are all *complex*. They lie on a *virtual* absolute quadric, whose canonical form is

$$(x^1)^2 + (x^2)^2 + (x^3)^2 + (x^4)^2 = 0.$$

Its matrix Q is simply the unit 4×4 matrix I. Clearly, every collineation whose matrix satisfies

$$SS^T = I$$

is an isometry. The isometries constitute the *orthogonal group* $O(4, R)$. In terms of the *complex* coordinates $(y^1\ y^2\ y^3\ y^4) = (x^4 + ix^3\ x^1 + ix^2\ -x^1 + ix^2\ x^4 - ix^3)$ the equation of the absolute becomes $y^1 y^4 = y^2 y^3$ which is just

$$|Y| = 0, \quad Y = \begin{pmatrix} y^1 & y^2 \\ y^3 & y^4 \end{pmatrix}$$

But now this matrix is not real—without loss of generality it can, for finite points, be taken to be unimodular. It is then *unitary* (its inverse is its *Hermitian conjugate*, the complex conjugate of its transpose: $Y\bar{Y}^T = YY^\dagger = I$). To preserve the reality of the coordinates $(x^1\ x^2\ x^3\ x^4)$ of elliptic space, the unitary property of Y must be preserved. For

$$Y \to LYR$$

to be an isometry the left and right translations must also be unitary,

$$L^{-1} = L^\dagger, \qquad R^{-1} = R^\dagger$$

L can be taken to be unimodular, so that the left translations constitute the special unitary group $SU(2)$. The right translations, which commute with them, constitute another special unitary group. The group of isometries can be expressed as $SU(2) \otimes SU(2) \sim O(4, R)$.

We ought at last to say something about the isometries for a metric space with a singular absolute. And what better example than the familiar Euclidean geometry $\{0--\}$? The absolute is then given, as we have seen, by two equations

$$x^4 = 0, \qquad (x^1)^2 + (x^2)^2 + (x^3)^2 = 0.$$

As is easily verified, the collineations that leave these equations invariant have the form

$$S = \begin{pmatrix} s_1^1 & s_2^1 & s_3^1 & a^1 \\ s_1^2 & s_2^2 & s_3^2 & a^2 \\ s_1^3 & s_2^3 & s_3^3 & a^3 \\ 0 & 0 & 0 & 1 \end{pmatrix}$$

4.18 Isometries

where the (real) 3×3 submatrix s is orthogonal ($ss^T = 1$). Denote the vector $(a_1\ a_2\ a_3)$, written as a column, by **a**. Every isometry may be classified as a *translation* ($s = 0$), a glide reflection ($|s| = -1$, $s\mathbf{a} = \mathbf{a}$, $\mathbf{a} \neq 0$), a reflection ($|s| = -1$, $s\mathbf{a} \neq \mathbf{a}$ or $\mathbf{a} = 0$), a screw transformation ($|s| = 1$, $s\mathbf{a} = \mathbf{a}$, $\mathbf{a} \neq 0$), or a rotation ($|s| = 1$, $s\mathbf{a} \neq \mathbf{a}$ or $\mathbf{a} = 0$). The usual Cartesian coordinates in Euclidean space are the inhomogeneous coordinates $(X\ Y\ Z) = (x^1/x^4\ x^2/x^4\ x^3/x^4)$.

We can proceed as we did for elliptic space, defining the complex coordinates $(y^1\ y^2\ y^3\ y^4) = (x^4 + x^1\ ix^2 + x^3\ -ix^2 + x^3\ x^4 - x^1)$ and considering the transformations

$$Y \to LYR$$

of the matrix

$$Y = \begin{pmatrix} y^1 & y^2 \\ y^3 & y^4 \end{pmatrix}.$$

In this case it is Hermitian ($Y = Y^\dagger$). But now the absolute is given not just by $|Y| = 0$, but by

$$|Y| = 0, \quad \operatorname{trace} Y = 0.$$

The transformations that preserve these equations while keeping Y Hermitian are

$$Y \to LYL^\dagger + (\operatorname{trace} Y)A$$

where A is any traceless Hermitian 2×2 matrix. This second term corresponds to the Euclidean translations, with

$$A = \frac{1}{2} \begin{pmatrix} a^1 & ia^2 + a^3 \\ -ia^2 + a^3 & -a^1 \end{pmatrix}.$$

The matrices L must be *unitary* matrices ($LL^\dagger = I$) so that trace Y ($= 2x^4$) will be invariant. Again, we may take them to be unimodular, $|L| = 1$, so the group of isometries consists of the translations **a** and the *special unitary group* SU(2). We have a group isomorphism $O(3, R) \sim SU(2)$.

Chapter 5
Cubic Figures

Abstract This chapter deals with plane cubic curves, 'twisted' cubic curves in three dimensions, cubic surfaces, and the various canonical and parametric forms of these constructs. The theory of the 27 lines on a general cubic surface is presented in detail.

5.1 Plane Cubic Curves

A *plane algebraic curve* is defined by a homogeneous polynomial equation in the homogeneous coordinates $(X\ Y\ Z)$. An investigation of the properties of *plane cubic curves* serves to illustrate some of the general features of higher-order plane algebraic curves, such as *nodes*, *cusps* and *inflections*, that are not revealed by a discussion of conics.

Writing the set of coordinates $(X\ Y\ Z)$ for a general point as X or as $(x^1\ x^2\ x^3)$ and employing the summation convention, the equation of a plane cubic curve F is $F(X) = 0$, where

$$F(X) = F_{ijk} x^i x^j x^k.$$

We shall assume F is a *real* cubic curve—the coefficients F_{ijk} are all real numbers. (There are ten of them because of the symmetries $F_{ijk} = F_{ikj}$ etc.) However, as in the case of conics, it is often illuminating to take into consideration points with complex coordinates.

It is convenient to also define

$$L_i(X) = F_{ijk} x^j x^k, \qquad Q_{ij}(X) = F_{ijk} x^k.$$

Tangents, polarity, affine specializations, etc can be defined as for quadratic curves (conics). For example, every point A has a *polar line* with respect to a general cubic. It is the line $L(A)$ with homogeneous dual coordinates $[L_1(A)\ L_2(A)\ L_3(A)]$. A point A lies on its polar line if and only if it lies on the cubic curve. In that case the polar line is a *tangent* to the curve, touching it at A. But now there is also another kind of polarity:

Every point A of the plane also has a *polar conic*, whose matrix $Q(A)$ is given by $Q_{ij}(A)$, and a point A lies on its polar conic if and only if it lies on the curve.

Fig. 5.1 A cubic curve with a node, and another with a cusp

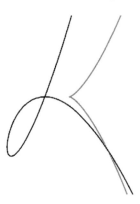

A line AB cuts a plane cubic curve F in the *three* points $A + \lambda B$ where λ is a root of $F(A + \lambda B) = 0$; that is,

$$\lambda^3 F(B) + 3\lambda^2 B^T Q(A) B + 3\lambda A^T Q(B) A + F(A) = 0$$

which can also be written as

$$\lambda^3 F(B) + 3\lambda^2 L(B) A + 3\lambda L(A) B + F(A) = 0.$$

If two of the roots are coincident the line AB is a *tangent* to the curve. For a point A on the curve, $F(A) = 0$ and the equation then has a root $\lambda = 0$. If $L(A)B = 0$ it has two coincident roots $\lambda = 0$. The line AB is then the line $L(A)$ and is the tangent to the curve at A. However, if $L(A) = 0$ a unique tangent at A cannot be defined and A is a *double point* or *node* of the cubic curve. If A is a node and B is any point on the polar conic of A ($B^T Q(A) B = 0$), the equation has *three* coincident roots at $\lambda = 0$. The polar conic of A then contains the whole line AB and is therefore degenerate, consisting of two lines.

> *The polar conic of a node is degenerate, consisting of the two tangents at the node.*

(Figure 5.1.) A node is called a *cusp* if these two tangents coincide.

Suppose a plane cubic curve has two nodes, at two points A and B. Then $F(A) = F(B) = L(A)B = L(B)A = 0$. Hence $F(A + \lambda B) = 0$ for all λ; the curve contains the whole line AB. Hence:

> *A plane cubic curve with two nodes is degenerate, consisting of a line and a conic.*

(The conic may also, of course, be degenerate and then the cubic curve is simply three lines.)

A point A on a plane algebraic curve is a *triple point* if *every* line through A cuts the curve in three coincident points. In the case of a plane cubic curve, this means $F(A + \lambda B) = \lambda^3 F(B)$ for any point B, and hence if B is on F, so is every point on AB:

> *A plane cubic curve with a triple point is degenerate, consisting of three concurrent lines.*

5.2 Nine Points

There is a plane cubic curve through any nine given points.

If the nine points are $(X_1\ Y_1\ Z_1), (X_2\ Y_2\ Z_2), \ldots, (X_9\ Y_9\ Z_9)$, its equation is

$$\begin{vmatrix} X^3 & Y^3 & Z^3 & Y^2Z & Z^2Y & Z^2X & X^2Z & X^2Y & Y^2X & XYZ \\ X_1^3 & Y_1^3 & Z_1^3 & Y_1^2Z_1 & Z_1^2Y_1 & Z_1^2X_1 & X_1^2Z_1 & X_1^2Y_1 & Y_1^2X_1 & X_1Y_1Z_1 \\ X_2^3 & Y_2^3 & Z_2^3 & Y_2^2Z_2 & Z_2^2Y_2 & Z_2^2X_2 & X_2^2Z_2 & X_2^2Y_2 & Y_2^2X_2 & X_2Y_2Z_2 \\ X_3^3 & Y_3^3 & Z_3^3 & Y_3^2Z_3 & Z_3^2Y_3 & Z_3^2X_3 & X_3^2Z_3 & X_3^2Y_3 & Y_3^2X_3 & X_3Y_3Z_3 \\ X_4^3 & Y_4^3 & Z_4^3 & Y_4^2Z_4 & Z_4^2Y_4 & Z_4^2X_4 & X_4^2Z_4 & X_4^2Y_4 & Y_4^2X_4 & X_4Y_4Z_4 \\ X_5^3 & Y_5^3 & Z_5^3 & Y_5^2Z_5 & Z_5^2Y_5 & Z_5^2X_5 & X_5^2Z_5 & X_5^2Y_5 & Y_5^2X_5 & X_5Y_5Z_5 \\ X_6^3 & Y_6^3 & Z_6^3 & Y_6^2Z_6 & Z_6^2Y_6 & Z_6^2X_6 & X_6^2Z_6 & X_6^2Y_6 & Y_6^2X_6 & X_6Y_6Z_6 \\ X_7^3 & Y_7^3 & Z_7^3 & Y_7^2Z_7 & Z_7^2Y_7 & Z_7^2X_7 & X_7^2Z_7 & X_7^2Y_7 & Y_7^2X_7 & X_7Y_7Z_7 \\ X_8^3 & Y_8^3 & Z_8^3 & Y_8^2Z_8 & Z_8^2Y_8 & Z_8^2X_8 & X_8^2Z_8 & X_8^2Y_8 & Y_8^2X_8 & X_8Y_8Z_8 \\ X_9^3 & Y_9^3 & Z_9^3 & Y_9^2Z_9 & Z_9^2Y_9 & Z_9^2X_9 & X_9^2Z_9 & X_9^2Y_9 & Y_9^2X_9 & X_9Y_9Z_9 \end{vmatrix} = 0.$$

This cubic equation in $(X\ Y\ Z)$ is, of course, satisfied by any of the nine points because if the first row of a matrix is replaced by any other row, the determinant vanishes.

5.3 A Canonical Form for a Plane Cubic Curve

There are ten homogeneous coefficients F_{ijk}. If A, B and C are linear expressions in the homogeneous coordinates x^i, there will be ten homogeneous parameters in the expression

$$A^3 + B^3 + C^3 - 3\mu ABC$$

so we might expect that, in general, a given cubic polynomial F(X) can be expressed in this form. Suppose then that the equation of a cubic curve is given in the form

$$A^3 + B^3 + C^3 - 3\mu ABC = 0.$$

If none of the linear expressions A, B and C is identically zero and if the three lines $A = 0$, $B = 0$ and $C = 0$ are not concurrent, these three lines can be chosen as the sides of the reference triangle. The equation of the curve F is then

$$\left(x^1\right)^3 + \left(x^2\right)^3 + \left(x^3\right)^3 - 3\mu x^1 x^2 x^3 = 0,$$

and so we have a *canonical form*

$$X^3 + Y^3 + Z^3 - 3\mu XYZ = 0$$

for a plane cubic curve. As we might expect from this derivation, there are cubic curves that cannot be transformed to this canonical form. In fact, a cubic curve that has a node cannot be cast in this canonical form unless it is degenerate, consisting of three lines. At a node $F_i = 0$ and therefore we would have $X^2 = \mu YZ$, $Y^2 = \mu ZX$, $Z^2 = \mu XY$. Multiplying these three conditions, respectively, by X, Y and Z shows us that, at a node, $XYZ \neq 0$. Multiplying the three conditions together then gives $\mu^3 = 1$. Hence, if a plane cubic curve with canonical form $X^3 + Y^3 + Z^3 - 3\mu XYZ = 0$ has a node, μ is a *cube root of unity* $(1, \omega = (1 + i\sqrt{3})/2$ or $\bar{\omega} = (1 - i\sqrt{3})/2)$. The three identities

$$X^3 + Y^3 + Z^3 - 3XYZ = (X + Y + Z)(X + \omega Y + \bar{\omega}Z)(X + \bar{\omega}Y + \omega Z)$$

$$X^3 + Y^3 + Z^3 - 3\omega XYZ = \bar{\omega}(\omega X + Y + Z)(X + \omega Y + Z)(X + Y + \omega Z)$$

$$X^3 + Y^3 + Z^3 - 3\bar{\omega}XYZ = \omega(\bar{\omega}X + Y + Z)(X + \bar{\omega}Y + Z)(X + Y + \bar{\omega}Z),$$

which are consequences of $1 + \omega + \bar{\omega} = 0$ and $\omega\bar{\omega} = 1$, show that a cubic with a node cannot be cast in the canonical form $X^3 + Y^3 + Z^3 - 3\mu XYZ = 0$ *unless* it is degenerate—consisting of three (imaginary) lines.

5.4 Parametric Form

A plane cubic curve can be defined in terms of a parameter θ, as the set of points

$$\begin{pmatrix} X(\theta) & Y(\theta) & Z(\theta) \end{pmatrix},$$

where X, Y and Z are cubic polynomials in the variable θ. There are 12 coefficients in these three polynomials. If the parameter is changed to $\theta' = \alpha\theta + \beta$ (α and β arbitrary constants), the points $(X\ Y\ Z)$ on the curve are still specified by cubic polynomials, and the curve is not altered. Also, since the homogeneous coordinates $(X\ Y\ Z)$ are defined only up to an irrelevant overall factor, we have in this prescription essentially $12 - 3 = 9$ independent coefficients.

Substituting $x^1 = X(\theta)$, $x^2 = Y(\theta)$, $x^3 = Z(\theta)$ into $F_{ijk}x^i x^j x^k = 0$ gives a polynomial of ninth degree in θ. Equating coefficients of $1, \theta, \theta^2, \ldots, \theta^9$ gives ten linear homogeneous equations for the ten homogeneous coefficients F_{ijk}, which can be solved for their ratios. Conversely, given the F_{ijk}, the equation $F_{ijk}x^i(\theta)x^j(\theta)x^k(\theta) = 0$ can be solved for the coefficients in the three cubics $x^i(\theta)$. Therefore, the parametric definition of a cubic curve, and the definition in terms of a homogeneous equation $F_{ijk}x^i x^j x^k = 0$, are *equivalent*.

The parametric form provides a very simple way to see that:

> *Two general plane cubics have nine common points.*

Take one of the curves in parametric form and substitute into the equation $F = 0$ of the other. We get an equation of ninth degree in θ whose roots determine the intersections of the two curves.

5.5 Inflections

A tangent to a plane cubic curve F at a point that is *not* a node is an *inflectional tangent* if its three points of intersection with the curve all coincide, and its point of contact with F is then an *inflection* of F. A point A is an inflection of F if $F(A) = 0$, $L(A) \neq 0$, and $B^T Q(A) B = 0$ for every point B for which $L(A)B = 0$. Hence:

> The polar conic of an inflectional point on a cubic is degenerate, and contains the inflectional tangent.

Let us consider *all* the points of the plane whose polar conics are degenerate. A conic is degenerate (consisting of a pair of lines) if its matrix Q is singular. The polar conic of a point X is therefore degenerate if and only if

$$\begin{vmatrix} Q_{11}(X) & Q_{12}(X) & Q_{13}(X) \\ Q_{21}(X) & Q_{22}(X) & Q_{23}(X) \\ Q_{31}(X) & Q_{32}(X) & Q_{33}(X) \end{vmatrix} = 0.$$

This is the equation of another plane cubic curve, called the *Hessian* of F. The Hessian of a plane cubic curve is the locus of all points whose polar conics are degenerate.

> The inflections of a plane cubic curve F are the intersections of F with its Hessian that are not nodes of F.

Let F be a plane cubic curve given in the canonical form $X^3 + Y^3 + Z^3 - 3\mu XYZ = 0$. Its Hessian H is found to be

$$X^3 + Y^3 + Z^3 - 3\nu XYZ = 0, \quad \text{where } 3\nu = \mu - 4/\mu^2.$$

The nine points

$$\begin{array}{ccc} (0 \quad 1 \quad -1) & (-1 \quad 0 \quad 1) & (1 \quad -1 \quad 0) \\ (0 \quad 1 \quad -\omega) & (-\omega \quad 0 \quad 1) & (1 \quad -\omega \quad 0) \\ (0 \quad 1 \quad -\bar{\omega}) & (-\bar{\omega} \quad 0 \quad 1) & (1 \quad -\bar{\omega} \quad 0) \end{array}$$

satisfy $X^3 + Y^3 + Z^3 = 0$ and $XYZ = 0$ and therefore lie on both F and H. Since two cubics can have no more than nine common points, F has no other inflections. Moreover, every cubic of the form $F + \rho H = 0$ passes through these same nine points. Since there are three values of μ for a given value of ν, every cubic of this *pencil* of cubics is a Hessian for three others.

It is easily verified that the nine points lie in threes on the sides of four triangles:

> A general plane cubic has nine inflections forming a configuration $(9_4 12_3)$.

This is called *Jacobi's configuration*. A diagram using 12 straight lines cannot be given because the points cannot all be real, but Fig. 5.2 indicates the incidences by representing lines as circles or arcs. The nine points listed above have been named

$$\begin{array}{ccc} 1 & 2 & 3 \\ 4 & 5 & 6 \\ 7 & 8 & 9 \end{array}$$

Fig. 5.2 Jacobi's configuration

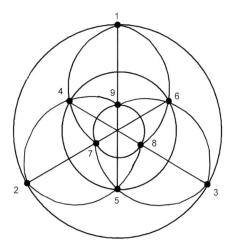

The 12 lines correspond to the rows, columns and 'terms in the determinantal expansion' of this array.

5.6 Cubics in Affine Geometry

Recall how a non-degenerate conic in affine geometry can be classified as an ellipse, a parabola or a hyperbola, according to how it is related to the 'line at infinity'. For cubic curves, *linear asymptotes* can be defined as for conics. They are the tangents at the points of intersection of the curve with the line at infinity—i.e. the *polar lines* of those intersections. A general cubic has three linear asymptotes, which may be all real, or two of them may be conjugate complex lines. Figure 5.3 illustrates a cubic curve with three real linear asymptotes. A cubic curve that happens to be tangential to the line at infinity has a *parabolic asymptote*, which is the *polar conic* of the point of contact of the curve with the line at infinity (Fig. 5.4).

Fig. 5.3 A cubic curve with three real linear asymptotes

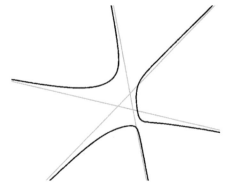

Fig. 5.4 A cubic curve with a linear and a parabolic asymptote

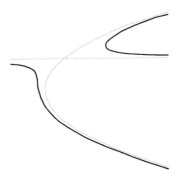

5.7 The Twisted Cubic

A curve in projective 3-space specified by

$$\begin{pmatrix} x^1 & x^2 & x^3 & x^4 \end{pmatrix} = (X \quad Y \quad Z \quad W) = \begin{pmatrix} X(\theta) & Y(\theta) & Z(\theta) & W(\theta) \end{pmatrix},$$

where $X(\theta)$, $Y(\theta)$, $Z(\theta)$ and $W(\theta)$ are linearly independent cubic polynomials in the variable parameter θ, with real coefficients, is called a *twisted cubic curve*. This prescription can be written concisely in matrix notation. Let X denote $(x^1 \ x^2 \ x^3 \ x^4)$, written as a column, and let Θ denote $(\theta^3 \ \theta^2 \ \theta \ 1)$ written as a column. The above parametric equation is then simply

$$X = M\Theta,$$

where the 4×4 matrix M contains the coefficients of the four cubic polynomials. Since the four cubics are asserted to be linearly independent, L will be non-singular and we can apply the collineation $X \to L^{-1}X$. We have a *canonical parametric form*

$$(X \quad Y \quad Z \quad W) = \begin{pmatrix} \theta^3 & \theta^2 & \theta & 1 \end{pmatrix}$$

for any twisted cubic. Since any twisted cubic can be brought into this canonical form by a collineation, it follows that

All twisted cubic curves are projectively equivalent.

Employing the canonical form for a given twisted cubic, we see that it cuts a plane with homogeneous coordinates $[p_1 \ p_2 \ p_3 \ p_4]$ at values of θ given by the roots of

$$p_1\theta^3 + p_2\theta^2 + p_3\theta + p_4 = 0.$$

Hence:

A twisted cubic and a plane have three common points.

These points may be all real, in which case they might be all distinct or two of them may coincide or all three may coincide. Or there may be one real point and a pair of conjugate complex points.

The canonical form for a twisted cubic is analogous to the canonical form $(\theta^2 \ \theta \ 1)$ for a conic, so we might expect to be able to regard the parameter θ as a homographic parameter. This is in fact the case. Suppose we change the canonical parameter θ on a twisted cubic to

$$\theta' = \frac{\alpha\theta + \beta}{\gamma\theta + \delta}, \qquad \alpha\delta - \beta\gamma \neq 0.$$

The parametric equation for the curve is now

$$(X\ Y\ Z\ W) = \left((\delta\theta' - \beta)^3\ (\delta\theta' - \beta)^2(-\gamma\theta' + \alpha)\ (\delta\theta' - \beta)(-\gamma\theta' + \alpha)^2\ (-\gamma\theta' + \alpha)^3\right)$$

(we have multiplied throughout by $(-\gamma\theta' + \alpha)^3$). By changing the reference tetrahedron and unit point this can be brought into the new canonical form

$$(X\ Y\ Z\ W) = \left((\theta')^3\ (\theta')^2\ \theta'\ 1\right).$$

The curve is unchanged. Hence a twisted cubic curve can be treated as a *projective line* on which the parameter θ is a *homographic parameter*. In particular, any three points on the curve can be chosen to be the reference points ∞, 0, and 1.

5.8 Chords, Tangent Lines and Osculating Planes

A line joining two points of a curve in 3-space is a *chord*.

> A twisted cubic curve has a unique chord through every point that does not lie on the curve.

Proof Three points on the curve with canonical parameters μ, ν and θ lie on the plane whose equation is

$$\begin{vmatrix} X & Y & Z & W \\ \theta^3 & \theta^2 & \theta & 1 \\ \mu^3 & \mu^2 & \mu & 1 \\ \nu^3 & \nu^2 & \nu & 1 \end{vmatrix} = 0.$$

As θ is varied, keeping μ and ν fixed, we get all the planes through the chord joining the points μ and ν (the chord is the *axis* of this *pencil* of planes). Any two of these planes determine the chord—we can take, for example, $\theta = 0$ and $\theta = \infty$. The chord is given as the intersection of the two planes

$$\begin{vmatrix} X & Y & Z & W \\ 0 & 0 & 0 & 1 \\ \mu^3 & \mu^2 & \mu & 1 \\ \nu^3 & \nu^2 & \nu & 1 \end{vmatrix} = 0, \qquad \begin{vmatrix} X & Y & Z & W \\ 1 & 0 & 0 & 0 \\ \mu^3 & \mu^2 & \mu & 1 \\ \nu^3 & \nu^2 & \nu & 1 \end{vmatrix} = 0,$$

that is, the two planes $[1\ -\mu - \nu\ \mu\nu\ 0]$ and $[0\ 1\ -\mu - \nu\ \mu\nu]$. For a given point $(X\ Y\ Z\ W)$ the two equations

5.8 Chords, Tangent Lines and Osculating Planes

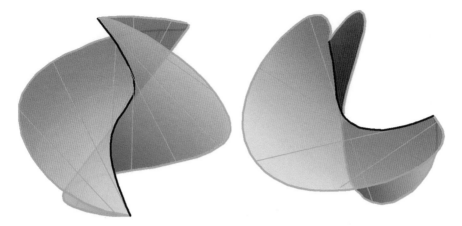

Fig. 5.5 The surface generated by the tangents to a twisted cubic

$$X - (\mu + \nu)Y + \mu\nu Z = 0$$
$$Y - (\mu + \nu)Z + \mu\nu W = 0$$

can be solved for $(\mu + \nu)$ and $\mu\nu$:

$$\mu + \nu = \alpha = (YZ - XW)/(Z^2 - YW)$$
$$\mu\nu = \beta = (Y^2 - ZX)/(Z^2 - YW).$$

μ and ν, the two points determining the unique chord through $(X\ Y\ Z\ W)$, are then the roots of the quadratic $\lambda^2 - \alpha\lambda + \beta = 0$.

A *tangent* is a chord that cuts the curve at a coincident pair of points. More specifically, the tangent at a point θ on a twisted cubic is the limit of the chord through θ and $(\theta + \varepsilon)$ as $\varepsilon \to 0$. Canonically, it is the line of points

$$(\theta^3 + 3\phi\theta^2 \quad \theta^2 + 2\phi\theta \quad \theta + \phi \quad 1)$$

where ϕ is a parameter on the line. Clearly, then, all the tangents to the twisted cubic lie on a *ruled surface*, parametrized by the pair of parameters θ and ϕ. Figure 5.5 illustrates two views of such a surface. The curve is a singularity of the surface called a *cuspidal edge*.

An *osculating plane* is a plane that touches the twisted cubic at three coincident points. The osculating plane $[p_1\ p_2\ p_3\ p_4]$ at the point θ must satisfy

$$p_1\lambda^3 + p_2\lambda^2 + p_3\lambda + p_4 = p_1(\lambda - \theta)^3.$$

Equating coefficients of λ^3, λ^2, λ and 1 reveals that the osculating plane at θ is

$$\begin{bmatrix}1 & -3\theta & 3\theta^2 & -\theta^3\end{bmatrix}.$$

Recalling that the tangent at θ is $(\theta^3 + 3\phi\theta^2\ \theta^2 + 2\phi\theta\ \theta + \phi\ 1)$, we see that:

> *The osculating plane at a point contains the tangent at that point*

(i.e., $(\theta^3 + 3\phi\theta^2) - 3\theta(\theta^2 + 2\phi\theta) + 3\theta^2(\theta + \phi) - \theta^3 = 0$).

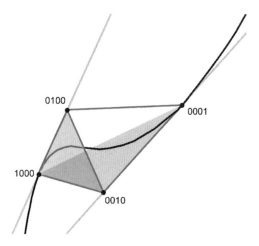

Fig. 5.6 The reference tetrahedron and the canonical form for a twisted cubic, showing two tangents and two osculating planes

An obvious consequence of all this is that, for a twisted cubic curve in canonical form, the three points $\theta = \infty, \theta = 0$ and $\theta = 1$ are two vertices of the reference tetrahedron and the unit point $-(1000), (0001)$ and (1111). The tangent at (0001) passes through (0010) and the osculating plane touching the curve at (0001) is $[1000]$. The tangent at (1000) passes through (0100) and the osculating plane touching the curve at (1000) is $[0001]$. The curve is related to the reference tetrahedron in the manner indicated in Fig. 5.6.

5.9 A Net of Quadrics

Any twisted cubic can be inscribed on a quadric. In fact,

Any twisted cubic curve is the common intersection of a threefold infinity of quadrics.

When a twisted cubic is expressed in canonical parametric form $(X\ Y\ Z\ W) = (\theta^3\ \theta^2\ \theta\ 1)$ we see that every point on it belongs to each of the three quadrics whose equations are

$$Q_1 = Y^2 - XZ = 0$$
$$Q_2 = YZ - XW = 0$$
$$Q_3 = Z^2 - YW = 0.$$

The curve therefore lies on every quadric of the *net* of quadrics whose equations have the form

$$\alpha Q_1 + \beta Q_2 + \gamma Q_3 = 0.$$

(In general, the intersection of a pair of quadrics is a quartic curve. In this case, any two quadrics of the net intersect in a degenerate quartic curve consisting of

5.10 Cubic Surfaces

Fig. 5.7 A twisted cubic (*white curve*) common to an elliptic cone (*black*), a hyperbolic paraboloid and a parabolic cylinder

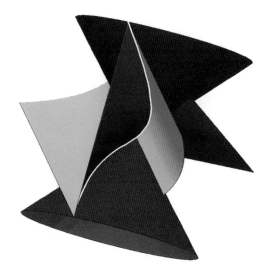

the twisted cubic and a common line. For example, the pairs $\{Q_2\ Q_3\}$, $\{Q_3\ Q_1\}$ and $\{Q_1\ Q_2\}$ contain, respectively, the common lines $Z = W = 0$, $Y = Z = 0$ and $X = Y = 0$.)

Figures 5.7 and 5.8 have been obtained by going to affine (or Euclidean) space, taking $W = 0$ as the line at infinity and choosing the inhomogeneous coordinates $(X\ Y\ Z) = (X/W\ Y/W\ Z/W)$. The three quadrics Q_1, Q_2 and Q_3 are then

$$Y^2 = XZ \quad \text{(an elliptic cone)}$$
$$YZ = X \quad \text{(a hyperbolic paraboloid)}$$
$$Z^2 = Y \quad \text{(a parabolic cylinder)}$$

This demonstrates that Q_1 and Q_3 are *quadric cones* in the projective space, with vertices at the points $\theta = 0$ and $\theta = \infty$ of the twisted cubic curve. They are degenerate members of the net of quadrics. Since any point on the curve can be taken to be the point $\theta = 0$, we deduce that *every* point on a twisted cubic is the vertex of a quadric cone that contains the entire curve, and we arrive at the result that:

> *The perspective projection of a twisted cubic curve on to any plane, from any point on the curve, is a conic.*

5.10 Cubic Surfaces

An *algebraic surface* in projective 3-space is defined by a homogeneous polynomial equation in the homogeneous coordinates $(X\ Y\ Z\ W)$.

Writing the set of coordinates $(X\ Y\ Z\ W)$ for a general point as X or as $(x^1\ x^2\ x^3\ x^4)$ and employing the summation convention, the equation of a cubic surface F is $F(X) = 0$ where

$$F(X) = F_{ijk} x^i x^j x^k.$$

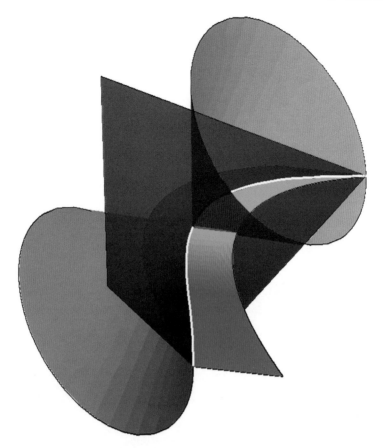

Fig. 5.8 Another view of the intersection of the three quadrics

We shall assume F is a *real* cubic surface—the coefficients F_{ijk} are all real numbers. There are *twenty* of them because of the symmetries $F_{ijk} = F_{ikj}$ etc., but the homogeneity means that $F(\lambda X) = 0$ specifies the same surface, so that a cubic surface is determined by 19 conditions. For example:

There is a cubic surface that contains any 19 given points.

The proof is entirely analogous to the proof that a plane cubic curve exists through any nine given points—the equation of the cubic surface is obtained as the vanishing of the determinant of a 20×20 matrix.

The section of a cubic surface by any plane is a cubic curve. (We can choose any plane to be $W = 0$ and then $F(X)$ becomes simply a cubic polynomial in the three homogeneous coordinates $(X\ Y\ Z)$.) The plane section may be degenerate, consisting of a conic and a line, or of three lines. A plane that intersects a cubic surface in three lines is a *tritangent plane*.

5.10 Cubic Surfaces

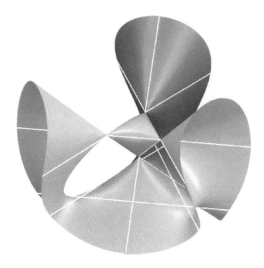

Fig. 5.9 A cubic surface with four nodes, containing nine real lines. Image from O Labs & D van Straten, *The Cubic Surface Homepage*, University of Mainz. Used with permission

Some of the properties of cubic surfaces are analogous to the properties of plane cubic curves. We can, for example, define

$$P_i(X) = F_{ijk}x^j x^k, \qquad Q_{ij}(X) = F_{ijk}x^k$$

and define a *polar plane* with respect to a cubic surface, for any point A. It is the plane P(A) with homogeneous dual coordinates [$P_1(A)$ $P_2(A)$ $P_3(A)$ $P_4(A)$]. A point A lies on its polar plane if and only if it lies on the surface. In that case the polar plane is a *tangent plane*, touching the surface at A. Every point A also has a *polar quadric*, whose matrix Q(A) is given by $Q_{ij}(A)$, and a point A lies on its polar quadric if and only if it lies on the surface.

A line AB cuts a cubic surface F in the *three* points $A + \lambda B$ where λ is a root of $F(A + \lambda B) = 0$; that is,

$$\lambda^3 F(B) + 3\lambda^2 B^T Q(A) B + 3\lambda A^T Q(B) A + F(A) = 0$$

which can also be written as

$$\lambda^3 F(B) + 3\lambda^2 P(B) A + 3\lambda P(A) B + F(A) = 0.$$

If two of the roots are coincident the line AB is a *tangent* to the surface. For a point A *on* the surface $F(A) = 0$, and the equation then has a root $\lambda = 0$. The condition for two coincident roots $\lambda = 0$ is $P(A)B = 0$. The line AB then lies in the tangent plane P(A) and is a tangent to the curve at A. The tangent plane at A contains all the tangent lines through A.

A *node* is a point A on the surface where $P(A) = 0$. It is, therefore, a point where a unique tangent plane cannot be defined (Fig. 5.9). If A is a node and B is any point on the polar quadric of A (i.e., $B^T Q(A) B = 0$), the equation $F(A + \lambda B)$ becomes simply $\lambda^3 F(B) = 0$, so this is the case of *three* coincident roots. The polar quadric contains the whole line AB. The polar quadric of a node A is therefore a quadric cone with vertex at A, generated by an infinity of tangent lines AB. It is the *tangent cone* at the node.

5.11 Canonical Forms for a Cubic Surface

There are four coefficients in a linear function of the four variables $(X\,Y\,Z\,W)$. If A_1, A_2, A_3, A_4 and A_5 are five linear functions of the four homogeneous coordinates $(X\,Y\,Z\,W)$, the homogeneous cubic equation

$$A_1^3 + A_2^3 + A_3^3 + A_4^3 + A_5^3 = 0$$

contains *twenty* coefficients and so we should expect to be able to express any homogeneous cubic equation $F(X) = 0$ (which also has 20 coefficients) in this form.

The set of five planes $A_i = 0$ is referred to as a *Sylvester pentahedron* if they are five *general* planes (i.e., no three have a common line). In that case, they can be chosen as the four reference planes and the unit plane and we get *Sylvester's canonical form* for a cubic surface,

$$\lambda_1 X^3 + \lambda_2 Y^3 + \lambda_3 Z^3 + \lambda_4 W^3 - \lambda_5 (X+Y+Z+W)^3 = 0,$$

or, equivalently,

$$\lambda_1 (x^1)^3 + \lambda_2 (x^2)^3 + \lambda_3 (x^3)^3 + \lambda_4 (x^4)^3 + \lambda_5 (x^5)^3 = 0,$$

where x^5 is defined by

$$x^1 + x^2 + x^3 + x^4 + x^5 = 0.$$

What we have here is essentially a description of a cubic surface in 3-space as a section by the unit hyperplane of a cubic hypersurface in 4-space.

Another interesting canonical form for a cubic surface expresses the equation of the surface as a determinant. Let A_1, A_2 and A_3 be three points. Let Π_1, Π_2 and Π_3 be three planes through A_1, let Π_4, Π_5 and Π_6 be three planes through A_2, and let Π_7, Π_8 and Π_9 be three planes through A_3. The set of all planes concurrent at a point is the dual of the set of all points in a plane, and hence is the 3-space *dual* of a projective plane. It is called a *star*. Any plane through A_1 has the form $\theta \Pi_1 + \phi \Pi_2 + \xi \Pi_3 = 0$ and can be specified by the three homogeneous parameters $(\theta\ \phi\ \xi)$. The locus of the point of intersection of the three corresponding planes

$$\theta \Pi_1 + \phi \Pi_2 + \xi \Pi_3 = 0$$
$$\theta \Pi_4 + \phi \Pi_5 + \xi \Pi_6 = 0$$
$$\theta \Pi_7 + \phi \Pi_8 + \xi \Pi_9 = 0,$$

as θ, ϕ and ξ vary has the equation

$$\begin{vmatrix} \Pi_1 & \Pi_2 & \Pi_3 \\ \Pi_4 & \Pi_5 & \Pi_6 \\ \Pi_7 & \Pi_8 & \Pi_9 \end{vmatrix} = 0.$$

The Π_i ($i = 1,\ldots,9$) here are linear functions of the homogeneous coordinates $(X\,Y\,Z\,W)$, so this is a *cubic surface*. The equation is the *determinantal canonical form* of a cubic surface. The Π_i are not unique for a given cubic surface, because the determinantal expression $|\Pi| = 0$ is unchanged by a transformation $\Pi \to A\Pi B$

of the 3 × 3 matrix Π, where A and B are two unimodular 3 × 3 matrices. (The transformations Π → ΠB keeps the three stars invariant and permutes the reference planes in each of the three stars; the transformations Π → AΠ correspond to a different set of three stars that produce the same cubic surface.) The P_i contains $4 \times 9 = 36$ coefficients and A and B each contain eight, so the number of independent coefficients in the canonical determinantal form $|\Pi| = 0$ is $36 - 2 \times 8 = 20$.

Hence:

A cubic surface is the locus of the point of intersection of corresponding planes of three projectively related stars.

A third possibility is to express the equation of a cubic surface in the form

$$ABC = DEF,$$

where A, B, C, D, E and F are linear functions of the coordinates. The expression ABC − DEF apparently contains 24 parameters, but is unchanged by multiplication of A, B, C, D, E and F, respectively, by factors a, b, c, d, e and f, provided $abc = def$, so there are essentially only 20 independent parameters, as required to specify a general cubic surface. In the most general case when no three of the six planes A = 0, B = 0, C = 0, D = 0, E = 0, F = 0 have a common line, we have an *associated trihedron pair* (see the discussion 'six planes in 3-space' and Figs. 3.5 and 3.6 therein). The surface contains the nine lines

$$\begin{array}{ccc} A \cdot D & B \cdot D & C \cdot D \\ A \cdot E & B \cdot E & C \cdot E \\ A \cdot F & B \cdot F & C \cdot F \end{array}$$

As we shall see later, a general cubic surface contains 27 lines, and can be expressed in the form ABC = DEF in 120 different ways.

5.12 Twenty-Seven Lines

As we have already noticed, cubic surfaces contain lines. This section is an examination of the configuration of the lines that are contained in a cubic surface.

Recall that in the determinantal canonical form of a general cubic surface every point on the surface is determined by three parameters $(\theta \ \phi \ \xi)$, as the point of intersection of three planes. As θ, ϕ and ξ are varied we would expect there to be exceptional situations where the three planes intersect in a common *line* rather than a common point. This line then lies entirely in the surface, and all its points are labeled by the same values of the three parameters. Characterizing a cubic surface in this way provides an important correspondence between the points on the surface and the points of *a projective plane* p, simply by regarding the homogeneous parameters $(\theta \ \phi \ \xi)$ as homogeneous coordinates for points on p. In general, to a point on p there corresponds a unique point on the surface, but there are *exceptional points* of p that correspond to *lines* in the surface. (As we shall see later, there are *six* of these exceptional points.)

Specifically, a cubic surface F is the locus of the points of intersection of three planes

$$\theta \Pi_1 + \phi \Pi_2 + \xi \Pi_3 = 0$$
$$\theta \Pi_4 + \phi \Pi_5 + \xi \Pi_6 = 0$$
$$\theta \Pi_7 + \phi \Pi_8 + \xi \Pi_9 = 0.$$

If the homogeneous coordinates of these three planes are $[p_1\ p_2\ p_3\ p_4]$. $[q_1\ q_2\ q_3\ q_4]$ and $[r_1\ r_2\ r_3\ r_4]$, these three equations are

$$p_1 X + p_2 Y + p_3 Z + p_4 W = 0$$
$$q_1 X + q_2 Y + q_3 Z + q_4 W = 0$$
$$r_1 X + r_2 Y + r_3 Z + r_4 W = 0$$

where, of course, p_i, q_i and r_i ($i = 1, \ldots, 4$) are homogeneous linear functions of the three variable parameters ($\theta\ \phi\ \xi$). The intersection of the cubic surface F by a *fixed* plane $[s_1\ s_2\ s_3\ s_4]$ is then given by

$$f(\theta\ \phi\ \xi) = \begin{vmatrix} p_1 & p_2 & p_3 & p_4 \\ q_1 & q_2 & q_3 & q_4 \\ r_1 & r_2 & r_3 & r_4 \\ s_1 & s_2 & s_3 & s_4 \end{vmatrix} = 0$$

where $f(\theta\ \phi\ \xi)$ is a homogeneous cubic polynomial in $(\theta\ \phi\ \xi)$. The cubic curve of intersection of F by a plane therefore corresponds to a cubic curve in p. Since every plane $[s_1\ s_2\ s_3\ s_4]$ cuts every line on the cubic surface F, every such cubic curve f in the plane p contains all the exceptional points. Two different planes have *three* points in common with F (the three points where their common line cuts F). The two corresponding cubic curves in p have nine common points. Hence, in the most general case, the number of exceptional point is *six*. Call them

$$A_1 \quad A_2 \quad A_3 \quad A_4 \quad A_5 \quad A_6$$

They correspond to six *lines* $a_1\ a_2\ a_3\ a_4\ a_5\ a_6$ on F.

Since we are considering only the most general case we can assert that no three of the exceptional points are collinear, so there is a unique conic through any five of them.

There are *six conics*

$$B_1 \quad B_2 \quad B_3 \quad B_4 \quad B_5 \quad B_6$$

that pass through five exceptional points (B_1 through $A_2\ A_3\ A_4\ A_5\ A_6$ etc.), and there are 15 *lines*

$$C_{12} \quad C_{13} \quad C_{14} \quad C_{15} \quad C_{16} \quad C_{23} \quad C_{24} \quad C_{25} \quad C_{26} \quad C_{34} \quad C_{35} \quad C_{36}$$
$$C_{45} \quad C_{46} \quad C_{56}$$

that pass through two exceptional points (C_{12} through A_1 and A_2 etc.).

5.12 Twenty-Seven Lines

Now consider a plane that contains a line in F that is *not* one of the six lines a_i. It cuts F in a *degenerate* cubic curve (i.e., a conic and a line), which has two *double points* (nodes). This corresponds, in p, to a cubic curve f with two nodes, which is therefore also degenerate. There are two possibilities: either f consists of the unique conic through *five* exceptional points and a line through the sixth, or it consists of the unique line through *two* exceptional points and a conic through the other four.

Thus we have a correspondence between the lines on the cubic surface F and the exceptional points, the conics through five of them, and the lines through two of them, in p. Thus we have $6 + 6 + 15 = 27$ lines on F. In an obvious notation, these lines may be labeled a_i, b_i and c_{ij}.

No two of the lines a_i in F corresponding to the exceptional points intersect each other, since that would imply two values of $(\theta \, \phi \, \xi)$ for their point of intersection. The conic B_1 contains $A_2 \, A_3 \, A_4 \, A_5 \, A_6$ but not A_1; hence the line b_1 intersects the lines $a_2 \, a_3 \, a_4 \, a_5 \, a_6$ but not a_1. The lines b_1 and b_2 do not intersect, because the plane through them would correspond to a cubic curve in P containing two conics, which is nonsense. Hence

The 12 lines a_i and b_i on the cubic surface F constitute a double-six.

Now note that C_{12} contains A_1 and A_2. C_{12} and B_1 together constitute a degenerate cubic through all six of the exceptional points, as do C_{12} and B_2. Also C_{12}, C_{34} and C_{56} constitute a degenerate cubic through the six exceptional points. Hence

$$c_{12} \text{ intersects } a_1 \quad a_2 \quad b_1 \quad b_2 \quad c_{34} \quad c_{35} \quad c_{36} \quad c_{45} \quad c_{46} \quad c_{56}$$

(and all other statements obtained from this one by permutations of 123456). Therefore:

The 15 lines c_{ij} are the associated lines of double-six a_i, b_i.

The 45 tritangent planes are those that contain the triangles such as $a_1 \, b_2 \, c_{12}$ (30 triangles obtained by permuting 123456) and $c_{12} \, c_{34} \, c_{56}$ (15 triangles):

A general cubic surface contains 27 lines and has 45 tritangent planes, constituting a configuration $(27_5 45_3)$ *of lines and planes, determined by any double-six of the lines.*

There are even real cubic surfaces on which all the 27 lines are *real*. Figure 5.10 shows a model of the *Clebsch diagonal surface*

$$x_1^3 + x_2^3 + x_3^3 + x_4^3 + x_5^3 = x_1 + x_2 + x_3 + x_4 + x_5 = 0$$

on which all the 27 lines are real and distinct. Notice that there are points where three of the lines intersect—a consequence of the high symmetry of the Clebsch surface—so this is not the most general case. These are *Eckhardt points* and there are ten of them. Real cubic surfaces exist, with less symmetry than this one, on which all the 135 points of the configuration

135	2	3
10	27	5
27	3	45

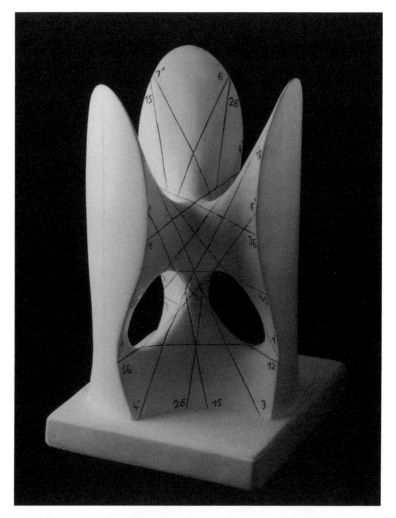

Fig. 5.10 The Clebsch surface. Photo from G Fischer, *Mathematische Modelle*, Vieweg 1986 (Chap. 2, Fig. 10, p. 13). Used with permission

are real and distinct. If we think of the Clebsch surface as a section by the unit hyperplane of a cubic hypersurface in 4-space, it is clearly invariant under a group S_5 of *collineations* given by permutations of the five homogeneous coordinates. It is not difficult to verify that the line joining $(1\ -1\ 0\ 0\ 0)$ and $(0\ 0\ 1\ -1\ 0)$ lies on this surface and, of course, so do all lines obtained from this one by permuting the coordinates (15 lines). The (real!) line joining $(1\ \xi\ \xi^2\ \xi^3\ \xi^4)$ to its complex conjugate, where ξ is the fifth root of unity $e^{2i\pi/5}$, and all lines obtained from this one by permutation of the five coordinates, also lie on the surface (12 lines); $15 + 12 = 27$.

Chapter 6
Quartic Figures

Abstract A few special quartic surfaces are introduced: the Hessian of a cubic surface, 'desmic' surfaces, and Kummer's quartic surface with its 16 nodes and 16 conics.

6.1 Algebraic Geometry

At this stage it becomes clear that the topics could go on and on. A homogeneous polynomial of degree n in $N + 1$ variables is called an *algebraic variety*. It is a hypersurface in N-dimensional projective space. The subject is vast, and leads into mathematics that is more like algebra than geometry.

An algebraic surface in 3-space can have *singularities* of various kinds. A large part of algebraic geometry is concerned with the elucidation of the different kinds of singularity and the classification of surfaces according to the number of singularities of the various kinds that a surface of given degree can have. As we have already seen in the case of cubic surfaces, the simplest kind of singularity is an isolated *node*—a point where the surface has no tangent plane, but is instead tangential to a quadric cone. If the tangent cone is degenerate, consisting of two planes, we get a *binode*. If the tangent cone degenerates to a single line we get a *cusp* (a sharp point on the surface—like a thorn...). A surface may also have *nodal curves*—curves along which the surface intersects itself. A cuspidal edge is a degenerate nodal curve (which we met already on the surface generated by the tangent lines of a twisted cubic—Fig. 5.5). Nodal curves can themselves have singularities, or can intersect, or can have end points. The topic is intricate and outside the scope of this book, but here in Fig. 6.1 are a few images illustrating some exotic possibilities (the coordinates in the caption are affine).

This book is meant to be 'elementary', so if you want to go deeper you will need to go to more advanced work. Nevertheless, there are fascinating things about quartic surfaces that it would be a pity to leave out. The few special quartic surfaces that we shall introduce in this section involve some particularly elegant mathematics, and they are related in surprising ways to the concepts we have met in earlier sections.

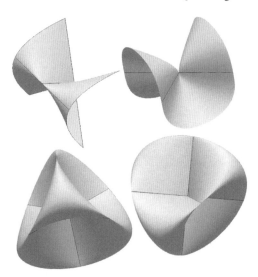

Fig. 6.1 Some surfaces with exotic singularities: 'Whitney's umbrella' $Z^2 - X^2Y = 0$; a surface with a 'binode', $Z^3 - X^2 + Y^2 = 0$; two views of Steiner's 'Roman' surface $X^2Y^2 + Z^2X^2 + Y^2Z^2 - XYZ = 0$

6.2 The Hessian of a Cubic Surface

The nodes of a cubic surface F are the points where $P_i = (1/3)\partial F/\partial x^i = 0$. If F is given by Sylvester's canonical form, this leads to

$$\lambda_1(x^1)^2 = \lambda_2(x^2)^2 = \lambda_3(x^3)^2 = \lambda_4(x^4)^2 = \lambda_5(x^5)^2$$

as the conditions to be satisfied at a node. This is the same as

$$x^1 : x^2 : x^3 : x^4 : x^5 = \frac{1}{\sqrt{\lambda_1}} : \frac{1}{\sqrt{\lambda_2}} : \frac{1}{\sqrt{\lambda_3}} : \frac{1}{\sqrt{\lambda_4}} : \frac{1}{\sqrt{\lambda_5}}$$

where each square root can take either sign $\pm$. This is compatible with $x^1 + x^2 + x^3 + x^4 + x^5 = 0$ only if

$$\frac{1}{\sqrt{\lambda_1}} + \frac{1}{\sqrt{\lambda_2}} + \frac{1}{\sqrt{\lambda_3}} + \frac{1}{\sqrt{\lambda_4}} + \frac{1}{\sqrt{\lambda_5}}$$

vanishes for at least one choice of the signs. Hence a cubic surface is devoid of singular points (nodes) only if this expression is not zero for any choice of the signs of the square roots. A cubic surface without nodes is a *non-singular* cubic surface.

We saw that a node of a cubic surface has a singular polar quadric—a tangent cone. Let us now consider *all* the points of the 3-space whose polar quadrics are singular. The polar quadric of a point X is singular if and only if

$$\begin{vmatrix} Q_{11} & Q_{12} & Q_{13} & Q_{14} \\ Q_{21} & Q_{22} & Q_{23} & Q_{24} \\ Q_{31} & Q_{32} & Q_{33} & Q_{34} \\ Q_{41} & Q_{42} & Q_{43} & Q_{44} \end{vmatrix} = 0,$$

6.2 The Hessian of a Cubic Surface

where $Q_{ij} = (1/6)\partial^2 F/\partial x^i \partial x^j$. This is the equation of a *quartic surface*, called the *Hessian* of the cubic surface F. The Hessian of a cubic surface is the locus of all points whose polar quadrics are singular. Employing the canonical form

$$A_1^3 + A_2^3 + A_3^3 + A_4^3 + A_5^3 = \lambda_1(x_1)^3 + \lambda_2(x_2)^3 + \lambda_3(x_3)^3 + \lambda_4(x_4)^3 + \lambda_5(x_5)^3$$
$$= 0$$

for the cubic surface, the equation of its Hessian turns out to be $H = 0$, where

$$H = \kappa_1 A^2 A^3 A^4 A^5 + \kappa_2 A^3 A^4 A^5 A^1 + \kappa_3 A^4 A^5 A^1 A^2 + \kappa_4 A^5 A^1 A^2 A^3$$
$$+ \kappa_5 A^1 A^2 A^3 A^4$$

where $\kappa_1 = (\lambda_1)^{-2/3}$ etc.

The Hessian is thus seen to contain the ten edges

$$\begin{array}{cccc} A_1 \cdot A_2 & A_1 \cdot A_3 & A_1 \cdot A_4 & A_1 \cdot A_5 \\ A_2 \cdot A_3 & A_2 \cdot A_4 & A_2 \cdot A_5 \\ A_3 \cdot A_4 & A_3 \cdot A_5 \\ A_4 \cdot A_5 \end{array}$$

of the pentahedron (for example, $H = 0$ and $A_1 = 0$ imply $A_2 A_3 A_4 A_5 = 0$ so the intersection of the Hessian with the plane $A_1 = 0$ consists of four pentahedron edges).

The polar plane of a point *with respect to the Hessian* is $[H_1 \ H_2 \ H_3 \ H_4]$ given by $H_i = \partial H/\partial x^i$. Choosing the five planes as reference planes and unit plane, we have

$$H_1 = (\kappa_1 - \kappa_5) A_2 A_3 A_4 + (\kappa_1 A_5 - \kappa_5 A_1)(A_2 A_3 + A_3 A_4 + A_4 A_2)$$
$$H_2 = (\kappa_2 - \kappa_5) A_1 A_3 A_4 + (\kappa_2 A_5 - \kappa_5 A_2)(A_3 A_4 + A_4 A_1 + A_1 A_3)$$
$$H_3 = (\kappa_3 - \kappa_5) A_1 A_2 A_4 + (\kappa_3 A_5 - \kappa_5 A_3)(A_4 A_1 + A_1 A_2 + A_2 A_4)$$
$$H_4 = (\kappa_4 - \kappa_5) A_1 A_2 A_3 + (\kappa_4 A_5 - \kappa_5 A_4)(A_1 A_2 + A_2 A_3 + A_3 A_1).$$

This then implies that the ten vertices of the pentahedron are nodes of the Hessian (for example, $A_1 = A_2 = A_3 = 0$ implies $H_1 = H_2 = H_3 = H_4 = 0$). Moreover, if the cubic surface F is non-singular, its Hessian can have no other nodes. First, observe that the Hessian has no node on a pentahedron edge, other than at the three vertices on that edge; for example, on the edge $A_1 \cdot A_2$ we have $H_1 = \kappa_1 A_3 A_4 A_5$ and $H_2 = H_3 = H_4 = 0$ so the only nodes on this edge are the three that lie at the three vertices on the edge. The points of the Hessian that do not lie on any plane of the pentahedron satisfy

$$\frac{\kappa_1}{A_1} + \frac{\kappa_2}{A_2} + \frac{\kappa_3}{A_3} + \frac{\kappa_4}{A_4} + \frac{\kappa_5}{A_5} = 0,$$

as can be seen by dividing

$$H = \kappa_1 A_2 A_3 A_4 A_5 + \kappa_2 A_3 A_4 A_5 A_1 + \kappa_3 A_4 A_5 A_1 A_2 + \kappa_4 A_5 A_1 A_2 A_3$$
$$+ \kappa_5 A_1 A_2 A_3 A_4$$

by $A_1 A_2 A_3 A_4 A_5$. Similarly, by dividing the expressions for the H_i by $A_1 A_2 A_3 A_4 A_5$ we see (after a bit of tricky algebra...) that the nodes of the Hessian that do not lie on pentahedron faces satisfy

$$\frac{\kappa_1}{A_1^2} = \frac{\kappa_2}{A_2^2} = \frac{\kappa_3}{A_3^2} = \frac{\kappa_4}{A_4^2} = \frac{\kappa_5}{A_5^2}.$$

This is the same as

$$\lambda_1(x^1)^2 = \lambda_2(x^2)^2 = \lambda_3(x^3)^2 = \lambda_4(x^4)^2 = \lambda_5(x^5)^2,$$

which, as we have seen, is satisfied only by the nodes of F. Therefore:

> A non-singular cubic surface that possesses a Sylvester canonical form determines a unique Desargues' (10_3) consisting of the nodes of its Hessian and the lines joining those nodes in threes, which lie entirely in the Hessian. The Sylvester pentahedron is then the set of five planes determined by this Desargues' configuration.

6.3 Desmic Surfaces

If the four equations for the faces of a tetrahedron are $A_1 = A_2 = A_3 = A_4 = 0$ then the quartic equation

$$A_1 A_2 A_3 A_4 = 0$$

is the equation for a degenerate quartic surface that consists simply of four planes— the four faces of the tetrahedron. Not a very interesting quartic surface! However, recall the canonical form

$$\begin{pmatrix} 0 & 1 & 0 & 1 \\ 1 & 0 & 1 & 0 \\ 0 & -1 & 0 & 1 \\ -1 & 0 & 1 & 0 \end{pmatrix} \begin{pmatrix} 0 & 1 & 0 & 1 \\ 0 & -1 & 0 & 1 \\ 1 & 0 & 1 & 0 \\ -1 & 0 & 1 & 0 \end{pmatrix} \begin{pmatrix} 1 & 0 & 1 & 0 \\ 0 & 1 & 0 & 1 \\ 0 & 1 & 0 & -1 \\ 1 & 0 & -1 & 0 \end{pmatrix}$$

for the three tetrahedra of a desmic system. Their vertices are given by the columns of these matrices. Their faces are given by the rows of

$$\begin{bmatrix} 0 & 1 & 0 & -1 \\ -1 & 0 & 1 & 0 \\ 0 & 1 & 0 & 1 \\ 1 & 0 & 1 & 0 \end{bmatrix} \begin{bmatrix} 0 & 0 & 1 & -1 \\ 1 & -1 & 0 & 0 \\ 0 & 0 & 1 & 1 \\ 1 & 1 & 0 & 0 \end{bmatrix} \begin{bmatrix} 1 & 0 & 0 & 1 \\ 0 & 1 & 1 & 0 \\ 1 & 0 & 0 & -1 \\ 0 & 1 & -1 & 0 \end{bmatrix},$$

6.3 Desmic Surfaces

so the three tetrahedra are given by the equations

$$\Delta_1 = A_1 A_2 A_3 A_4 = (Y - W)(Z - X)(Y + W)(Z + X)$$
$$= (Z^2 - X^2)(Y^2 - W^2) = 0$$
$$\Delta_2 = B_1 B_2 B_3 B_4 = (Z - W)(X - Y)(Z + W)(X + Y)$$
$$= (X^2 - Y^2)(Z^2 - W^2) = 0$$
$$\Delta_3 = C_1 C_2 C_3 C_4 = (X + W)(Y + Z)(X - W)(Y - Z)$$
$$= (Y^2 - Z^2)(X^2 - W^2) = 0.$$

From the identity

$$(Y^2 - Z^2)(X^2 - W^2) + (Z^2 - X^2)(Y^2 - W^2) + (X^2 - Y^2)(Z^2 - W^2) = 0$$

we see that

$$\Delta_1 + \Delta_2 + \Delta_3 = 0.$$

Since all desmic systems are projectively equivalent, it follows that if each of three expressions Δ_1, Δ_2 and Δ_3 is a product of four linear functions of $(X\ Y\ Z\ W)$ corresponding to the equations of the faces of three tetrahedra of a desmic system, then $\alpha \Delta_1 + \beta \Delta_2 + \gamma \Delta_3 = 0$ for some value of $(\alpha\ \beta\ \gamma)$, and by adjusting irrelevant factors of the homogeneous coordinates, the desmic system will satisfy

$$\Delta_1 + \Delta_2 + \Delta_3 = 0.$$

A *desmic surface* is a quartic surface given by an equation of the form

$$F = \mu \Delta_1 + \nu \Delta_2 + \rho \Delta_3 = 0.$$

The perspectivities of a desmic system (each pair of tetrahedra is perspective from each vertex of the third tetrahedron) imply the existence of 16 lines each of which is a line common to three faces. As is easily verified, they are the lines

$$\begin{array}{llll}
A_1 \cdot B_1 \cdot C_4 & A_2 \cdot B_1 \cdot C_3 & A_3 \cdot B_1 \cdot C_2 & A_4 \cdot B_1 \cdot C_1 \\
A_1 \cdot B_2 \cdot C_3 & A_2 \cdot B_2 \cdot C_4 & A_3 \cdot B_2 \cdot C_1 & A_4 \cdot B_2 \cdot C_2 \\
A_1 \cdot B_3 \cdot C_2 & A_2 \cdot B_3 \cdot C_1 & A_3 \cdot B_3 \cdot C_4 & A_4 \cdot B_3 \cdot C_3 \\
A_1 \cdot B_4 \cdot C_1 & A_2 \cdot B_4 \cdot C_2 & A_3 \cdot B_4 \cdot C_3 & A_4 \cdot B_4 \cdot C_4
\end{array}$$

These lines obviously all lie on the surface $F = 0$.

Recall now that any desmic system belongs to an *associated pair* of desmic systems. The two systems share the same 18 edges. The system associated with the one chosen for the above demonstration consists of the three tetrahedra

$$\begin{pmatrix} 1 & 0 & 0 & 0 \\ 0 & 1 & 0 & 0 \\ 0 & 0 & 1 & 0 \\ 0 & 0 & 0 & 1 \end{pmatrix} \begin{pmatrix} 1 & 1 & 1 & 1 \\ 1 & 1 & -1 & -1 \\ 1 & -1 & 1 & -1 \\ 1 & -1 & -1 & 1 \end{pmatrix} \begin{pmatrix} 1 & 1 & 1 & -1 \\ 1 & 1 & -1 & 1 \\ 1 & -1 & 1 & 1 \\ -1 & 1 & 1 & 1 \end{pmatrix}.$$

It is immediately obvious that all its 12 vertices lie on the surface F = 0. Moreover, they are *nodes* of the surface, because the four quantities

$$\partial F/\partial X = 2X\left[\mu\left(-Y^2 + W^2\right) + \nu\left(Z^2 - W^2\right) + \rho\left(Y^2 - Z^2\right)\right]$$
$$\partial F/\partial Y = 2Y\left[\mu\left(Z^2 - X^2\right) + \nu\left(-Z^2 + W^2\right) + \rho\left(X^2 - W^2\right)\right]$$
$$\partial F/\partial Z = 2Z\left[\mu\left(Y^2 - W^2\right) + \nu\left(X^2 - Y^2\right) + \rho\left(-X^2 + W^2\right)\right]$$
$$\partial F/\partial W = 2W\left[\mu\left(-Z^2 + X^2\right) + \nu\left(-X^2 + Y^2\right) + \rho\left(-Y^2 + Z^2\right)\right]$$

are obviously all zero for the 12 vertices of these three tetrahedra:

A general desmic surface contains 16 lines—the axes of the perspectivities of a desmic system—and has 12 nodes, located at the vertices of the three tetrahedra of the associated desmic system.

6.4 Kummer's Quartic Surface

There is a very large variety of quartic surfaces. The maximum number of isolated nodes that a quartic surface can have is 16. Quartic surfaces with this maximum number of isolated nodes were studied in detail by Kummer, and are called *Kummer surfaces*. Some of the nodes of a (real) Kummer surface may occur at complex points, so there are Kummer surfaces with fewer than 16 real nodes. These 16 nodes lie on 16 planes with six nodes on each plane and six planes through each node—a configuration (16_6) of points and planes.

Even more remarkable is that each of the 16 planes is tangential to the surface, *touching it at all the points of a conic*. (Incidentally—a tangent plane to a surface that is tangential to it at all points of a curve is called a trope.) So we have 16 conics on the Kummer surface each passing through six nodes. Figure 6.2 is a photo of a model demonstrating a Kummer surface in Euclidean space in which all the nodes are real and at finite points. It is taken from the frontispiece of Hudson's book *Kummer's Quartic Surface*, published in 1905. To count the conics and assure oneself that they each pass through six of the nodes, think of the three pairs of opposite outer portions (which ideally would extend to infinity) as 'deformed tetrahedra' (they each have four vertices at four of the nodes, like the other five pieces). These three pieces are cut in half by the plane at infinity. The surface is thus separated into eight pieces by its nodes, each piece containing four nodes and each node belonging to two pieces. The 16 quadrics in this Euclidean representation are four circles and 12 hyperbolae.

6.4 Kummer's Quartic Surface

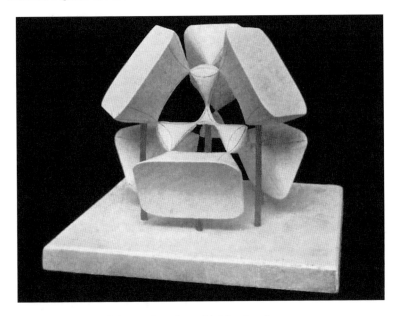

Fig. 6.2 A plaster model of Kummer's surface with 16 real nodes

We shall first think about the (16_6) of points and planes, and then relate it to the equation $F = 0$ of the surface, where F is a highly symmetrical homogeneous polynomial in $(X\,Y\,Z\,W)$. It is not difficult to convince oneself that the plane $[\alpha\,\beta\,\gamma\,\delta]$ in projective 3-space contains just six points obtainable by permuting the coordinates and inserting minus signs, namely

$$(\beta - \alpha \quad \delta - \gamma) \quad (\gamma \quad \delta - \alpha - \beta) \quad (\delta \quad \gamma - \beta - \alpha)$$
$$(\beta - \alpha - \delta \quad \gamma) \quad (\gamma - \delta - \alpha \quad \beta) \quad (\delta - \gamma \quad \beta - \alpha)$$

and, of course, we also have the dual statement that the point $(\alpha\,\beta\,\gamma\,\delta)$ lies on the six planes

$$[\beta - \alpha \quad \delta - \gamma] \quad [\gamma \quad \delta - \alpha - \beta] \quad [\delta \quad \gamma - \beta - \alpha]$$
$$[\beta - \alpha - \delta \quad \gamma] \quad [\gamma - \delta - \alpha \quad \beta] \quad [\delta - \gamma \quad \beta - \alpha]$$

Now we can begin with any one of these six planes instead of the one we started with; for example, the plane $[\beta - \alpha\,\delta - \gamma]$ contains the six points

$$(\alpha \quad \beta \quad \gamma \quad \delta) \quad (\delta - \gamma - \beta \quad \alpha) \quad (\gamma - \delta - \alpha \quad \beta)$$
$$(\alpha \quad \beta - \gamma - \delta) \quad (\delta \quad \gamma - \beta - \alpha) \quad (\gamma \quad \delta \quad \alpha \quad \beta)$$

And so on. We get 16 points in all, which may be named **1, 2, 3, ..., 16**:

1	(α	β	γ	δ)	**9**	(α	$-\beta$	γ	$-\delta$)
2	(β	α	δ	γ)	**10**	(β	$-\alpha$	δ	$-\gamma$)
3	(γ	δ	α	β)	**11**	(γ	$-\delta$	α	$-\beta$)
4	(δ	γ	β	α)	**12**	(δ	$-\gamma$	β	$-\alpha$)
5	(α	β	$-\gamma$	$-\delta$)	**13**	(α	$-\beta$	$-\gamma$	δ)
6	(β	$-\alpha$	$-\delta$	γ)	**14**	(β	α	$-\delta$	$-\gamma$)
7	(γ	δ	$-\alpha$	$-\beta$)	**15**	(γ	$-\delta$	$-\alpha$	β)
8	(δ	$-\gamma$	$-\beta$	α)	**16**	(δ	γ	$-\beta$	$-\alpha$)

And, of course, 16 planes which may be named *1, 2, 3, ..., 16* with the same corresponding sets of coordinates.

The incidence table, listing for each point the six planes it lies on, is

1	6 7 10 12 15 16	**9**	2 4 7 8 14 15
2	7 8 9 11 13 16	**10**	1 3 5 8 15 16
3	5 8 10 12 13 14	**11**	2 4 5 6 13 16
4	5 6 9 11 14 15	**12**	1 3 6 7 13 14
5	3 4 6 8 10 11	**13**	2 3 11 12 14 16
6	1 4 5 7 11 12	**14**	3 4 9 12 13 15
7	1 2 6 8 9 12	**15**	1 4 9 10 14 16
8	2 3 5 7 9 10	**16**	1 2 10 11 13 15

The table listing for each plane the six points that lie in it is the same, except that points and planes are interchanged: we have a self-dual (16_6) of points and planes. The order of its symmetry group is 192 and the symmetries are realizable as collineations.

There are many ways of expressing the Kummer quartic surface with 16 real nodes. A very elegant equation for the surface is

$$F = X^4 + Y^4 + Z^4 + W^4 - \left(Y^2Z^2 + Z^2X^2 + X^2Y^2 + X^2W^2 + Y^2W^2 + Z^2W^2\right)$$
$$= 0.$$

In this highly symmetrical form we see that it is invariant under a group of collineations consisting of the 24 permutations of $XYZW$ and changing signs of the coordinates (eight possible choices). Hence a group of collineations of order $24 \times 8 = 192$ that leave the surface—and its associated configuration (16_6) of points and planes—unchanged.

Now we have

$$\partial F/\partial X = 4X^3 - 2X\left(Y^2 + Z^2 + W^2\right)$$
$$\partial F/\partial Y = 4Y^3 - 2Y\left(Z^2 + W^2 + X^2\right)$$

6.4 Kummer's Quartic Surface

$$\partial F/\partial Z = 4Z^3 - 2Z(W^2 + X^2 + Y^2)$$
$$\partial F/\partial W = 4W^3 - 2W(X^2 + Y^2 + Z^2)$$

and it is easy to verify that the 16 points

1	(0	1	1	1)		9	(0	1	−1	1)
2	(1	0	1	1)		10	(1	0	1	−1)
3	(1	1	0	1)		11	(−1	1	0	1)
4	(1	1	1	0)		12	(1	−1	1	0)
5	(0	−1	1	1)		13	(0	1	1	−1)
6	(1	0	−1	1)		14	(−1	0	1	1)
7	(1	1	0	−1)		15	(1	−1	0	1)
8	(−1	1	1	0)		16	(1	1	−1	0)

all satisfy $F = \partial F/\partial X = \partial F/\partial Y = \partial F/\partial Z = \partial F/\partial W = 0$.

They are the 16 nodes. (This is just the particular case $\alpha\beta\gamma\delta = 0111$ of the (16_6) just described.) The polynomial F is obviously invariant under the symmetry group.

The six points in the plane [1110] are

$$(-1 \;\; 1 \;\; 0 \;\; 1) \quad (-1 \;\; 0 \;\; 1 \;\; 1) \quad (0 \;\; -1 \;\; 1 \;\; 1)$$
$$(1 \;\; -1 \;\; 0 \;\; 1) \quad (1 \;\; 0 \;\; -1 \;\; 1) \quad (0 \;\; 1 \;\; -1 \;\; 1)$$

which all lie on the quadric

$$2W^2 - (X^2 + Y^2 + Z^2) = 0$$

and therefore all lie on the conic of intersection of this quadric with the plane. Because of the symmetries of the situation, *all 16 planes contain a conic through the six nodes in that plane.*

To show that this conic

$$2W^2 - (X^2 + Y^2 + Z^2) = 0, \qquad X + Y + Z = 0$$

lies entirely on the quadric surface $F = 0$, it is convenient to write

$$x = X^2, \qquad y = Y^2, \qquad z = Z^2, \qquad w = W^2.$$

The equations for the conic are then

$$2w = x + y + z, \qquad \sqrt{x} + \sqrt{y} + \sqrt{z} = 0.$$

Squaring both sides of this first equation gives

$$4w^2 = x^2 + y^2 + z^2 + 2(yz + zx + xy),$$

the second equation leads to

$$x^2 + y^2 + z^2 = 2(yz + zx + xy)$$

and these two latter equations give us

$$2w^2 = x^2 + y^2 + z^2.$$

Now substitute into

$$\begin{aligned}F &= x^2 + y^2 + z^2 + w^2 - (yz + zx + xy + xw + yw + zw)\\ &= w^2 + (x^2 + y^2 + z^2) - w(x + y + z) - (yz + zx + xy).\end{aligned}$$

This expression vanishes. The conic therefore lies entirely on the quadric surface $F = 0$; because of the symmetries of the situation, so do all 16 of the conics.

An affine specialization of Kummer's surface $F = 0$ (different from the one shown in Fig. 6.2) can be obtained simply by taking $W = 0$ as the plane at infinity. Inhomogeneous affine coordinates are then $(X\ Y\ Z) = (X/W\ Y/W\ Z/W)$ and in terms of these coordinates the 16 nodes are

1	(0 1 1)		9	(0 1 −1)
2	(1 0 1)		10	(−1 0 −1)
3	(1 1 0)		11	(−1 1 0)
4	(1 1 1)		12	(1 −1 1)
5	(0 −1 1)		13	(0 −1 −1)
6	(1 0 −1)		14	(−1 0 1)
7	(−1 −1 0)		15	(1 −1 0)
8	(−1 1 1)		16	(1 1 −1)

The four on the bottom row are *at infinity*—for example, (1 1 1) denotes the point of concurrence of the family of parallel lines through the origin (000) and the point (111).

Interpreting $(X\ Y\ Z)$ as Cartesian coordinates in Euclidean space, we have 12 nodes at the mid-points of edges of a cube $(\pm 1\ \pm 1\ \pm 1)$ and four at the points at infinity on its four main diagonals (Fig. 6.3). There are four planes

4	5 6 9 11 14 15
8	2 3 5 7 9 10
12	1 3 6 7 13 14
16	1 2 10 11 13 15

like the one shown top right in the figure and 12 like the one shown lower right.

The equation of the surface in affine space is now

$$X^4 + Y^4 + Z^4 - X^2 - Y^2 - Z^2 - YZ - ZX - XY + 1 = 0$$

and Fig. 6.4 shows what it looks like when the coordinates $(X\ Y\ Z)$ are interpreted as Cartesian coordinates in Euclidean space. Each cut end is to be imagined as continuing to infinity where it converges on a node. Two opposite portions share a node at infinity. Again, we see that the surface is partitioned into eight pieces by its nodes, each piece having four nodes and each node common to two pieces.

6.4 Kummer's Quartic Surface

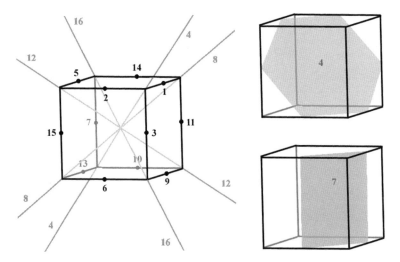

Fig. 6.3 The positions of the 16 nodes and the two kinds of planes

Fig. 6.4 A Euclidean version of Kummer's surface

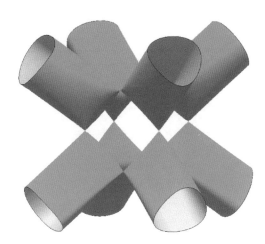

In this Euclidean representation four of the 16 conics on the surface are circles and 12 are hyperbolae. They are indicated in Fig. 6.5.

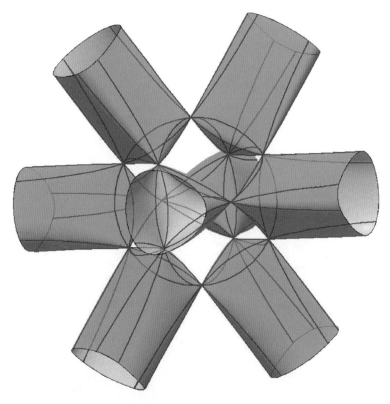

Fig. 6.5 The sixteen conics on a Kummer surface

Chapter 7
Finite Geometries

Abstract Galois fields and the finite Projective geometries that arise from them are introduced. The order of the collineation group for the finite geometry PG(N, q) is derived. Some particular cases with small values of N and q are described in detail. Some of the linear factional groups (homographies on finite projective lines) of special interest are presented and their relation to Steiner systems and Mathieu groups is explained. Finally, some remarkable special configurations are described, such as Coxeter's '12 points in PG(5, 3)', and '24 points in PG(11, 2)', which are geometrical realizations of Golay's ternary and binary codes.

7.1 Finite Geometries

So far, we have presumed coordinates of points, lines, planes, *et cetera*, have been presumed to be real numbers or complex numbers. There are other possibilities. So long as they are entities that can be added, subtracted, multiplied and divided according to the usual laws of arithmetic—so long as they belong to a *field*—the concepts and methods of analytic projective geometry remain valid. We are then dealing with abstract structures that may still be regarded as 'geometries'.

In particular, the field F_p of 'integers modulo p', where p is a prime, consists of p elements, denoted by the symbols $0, 1, 2, \ldots, p-1$. Addition and multiplication in F_p is just like addition and multiplication of the integers denoted by these same symbols, except that we retain only the remainder after dividing the result by p (for example $3 + 4 = 2$ modulo 5 and $6 \times 2 = 5$ modulo 7 ...).

A *point* in the N-dimensional projective geometry PG(N, F_p)—or PG(N, p)—is a set of homogeneous coordinates $(x^1 \, x^2 \, \ldots \, x^{N+1})$, each x^i belonging to F_p. There are p^{N+1} such $(N+1)$-tuples. But $(0 \, 0 \, \ldots \, 0)$ is excluded, and for all non-zero values of λ belonging to F_p the coordinates $\lambda(x^1 \, x^2 \, \ldots \, x^{N+1}) = (\lambda x^1 \, \lambda x^2 \, \ldots \, \lambda x^{N+1})$ all represent the same point. Hence the number of points in PG(N, p) is

$$\frac{p^{N+1} - 1}{p - 1} = p^N + p^{N-1} + \cdots + p + 1.$$

Fig. 7.1 The finite geometry PG(2, 2)

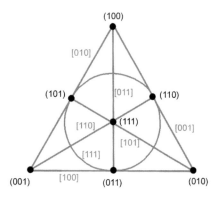

7.2 PG(2, 2)

F_2 has only two 'numbers' 0 and 1. Addition and multiplication 'modulo 2' is simply

$$0+0=0, \quad 0+1=1, \quad 1+1=0,$$
$$0\times 0=0, \quad 0\times 1=0, \quad 1\times 1=1.$$

The projective plane PG(2, 2) has seven points

(011) (101) (110) (100) (010) (001) (111)

and seven lines

[011] [101] [110] [100] [010] [001] [111].

There are three points on each line and three lines through each point. Figure 7.1 is a representation of the structure. It is, of course, the Fano configuration (7_3), Fig. 1.30.

In PG(2, 2) there are not even enough points and lines for Desargues' theorem or Pappus' theorem and the important concept of cross-ratio is absent—one needs four collinear points for that! It does, though, have quadrangles and quadrilaterals. Observe that the 'diagonal triangle' of a quadrangle degenerates to a single line and that of a quadrilateral degenerates to a single point.

The group of *collineations* (projective transformations), however, is of interest. It is the *linear fractional group* LF(3, 2), consisting of 3×3 unimodular matrices whose elements are integers modulo 2. Multiplying the columns of point coordinates $(X\ Y\ Z)$ on the left by any such matrix M and multiplying rows of line coordinates $[l\ m\ n]$ on the right by M^{-1} permutes the points and lines and preserves all the incidences $lX + mY + nZ = 0$. Since, in any projective 3-space, all quadrilaterals are projectively equivalent, the order of the group can be deduced by counting the number of different quadrilaterals in Fano's (7_3). This we have already done: the linear fractional group LF(3, 2) is the group of symmetries of Fano's (7_3), and is a group of order 168.

7.3 PG(2, 3)

The field F_3 consists of three symbols 0, 1, 2, added and multiplied using 'modulo 3' arithmetic. It is more convenient to write -1 instead of 2 (permissible because $-1 = 2 \bmod 3$). The geometry PG(2, 3) is a configuration (13_4) of points and lines. Name the 13 points **1, 2, ..., 13**, like this:

1	(1	0	0)	**8**	(−1	0	1)
2	(0	1	0)	**9**	(1	−1	0)
3	(0	0	1)	**10**	(−1	1	1)
4	(0	1	1)	**11**	(1	−1	1)
5	(1	0	1)	**12**	(1	1	−1)
6	(1	1	0)	**13**	(1	1	1)
7	(0	1	−1)				

and name the 13 lines *1, 2, ..., 13* analogously (*1* is [1 0 0] etc.).

The table of incidences, listing the four lines through each of the points, is easily found to be

1	**2**	**3**	**4**	**5**	**6**	**7**	**8**	**9**	**10**	**11**	**12**	**13**
2	*3*	*1*	*1*	*2*	*3*	*1*	*2*	*3*	*5*	*4*	*4*	*7*
3	*1*	*2*	*7*	*8*	*9*	*4*	*5*	*6*	*6*	*6*	*5*	*8*
4	*5*	*6*	*11*	*10*	*10*	*10*	*11*	*12*	*7*	*8*	*9*	*9*
7	*8*	*9*	*12*	*12*	*11*	*13*	*13*	*13*	*10*	*11*	*12*	*13*

and, of course, the table listing the four points on each of the lines is identical except that points and lines are interchanged.

The plane PG(2, 3) contains non-trivial Desargues' configurations. For instance, the two triangles **2 12 5** and **6 7 3** are perspective from **1**: by referring to the table we see that the three lines joining pairs of corresponding vertices are *3*, *4*, and *2*, which all contain **1**. The sides of the triangles are *12 8 5* and *1 9 10*; pairs of corresponding sides intersect at the points **4**, **13** and **10**, which all lie on the line *7*. All this is deducible from the incidence table.

A finite *affine* geometry is obtained by choosing a line at infinity. If the line *3* [001] is chosen, the four points at infinity are **1 2 6 9**. The remaining points and lines are the points and lines of a finite affine plane. It is a *Jacobi configuration* $(9_4\ 12_3)$ (see Fig. 5.2). The inhomogeneous coordinates $(X/Z\ Y/Z)$ of its nine points are

10	(−1	1)	**4**	(0	1)	**13**	(1	1)
8	(−1	0)	**3**	(0	0)	**5**	(1	0)
12	(−1	−1)	**7**	(0	−1)	**11**	(1	−1)

The 12 lines are given by the rows, columns and 'terms in the expansion of the determinant' of this 3×3 array. There are four sets of three 'parallel' lines.

7.4 PG(3, 2)

The three-dimensional projective geometry PG(3, 2) has 15 points, given by quadruples of modulo 2 integers, and 15 planes. Each plane, being a PG(2, 2), is a (7_3) of points and lines. The points and planes of PG(3, 2) therefore constitute a (15_7) of points and planes, and there are three points on every line. Duality then tells us that there are three planes through every line. The total number of lines is therefore $15 \times 7/3 = 35$. The structure of PG(3, 2) is

$$\begin{array}{|ccc|} \hline 15 & 7 & 7 \\ 3 & 35 & 3 \\ 7 & 7 & 15 \\ \hline \end{array}$$

7.5 Galois Fields

The integers modulo p satisfy the usual laws of arithmetic only if p is a prime. (Otherwise, there are 'divisors of zero', for example $2 \times 3 = 0$ mod 6.) There does exist, however, a finite field containing only q elements that can be added and multiplied, in a way that satisfies all the usual rules of arithmetic, for any q that is a *power* of a prime ($q = p^k$ with p a prime and k an integer). The elements of this field are not 'integers modulo q', they are essentially polynomials of degree less than or equal to $k - 1$ whose coefficients are integers modulo p. This is the *Galois field* GF(q) (which I shall henceforth write simply as F_q).

The elements of F_q, where $q = p^k$, are polynomials $a_0 + a_1 x + a_2 x^2 + \cdots + a_{k-1} x^{k-1}$. The coefficients $a_0, a_1, \ldots, a_{k-1}$ are integers modulo p; the nature of the variable x is *unspecified*. The polynomial $a_0 + a_1 x + a_2 x^2 + \cdots + a_{k-1} x^{k-1}$ can be denoted by the k-tuple $(a_0 \ a_1 \ \ldots \ a_{k-1})$.

Addition of the polynomials is obvious. Multiplication is not: we need a way of multiplying two polynomials of degree $k - 1$ so that the result is another polynomial of degree $k - 1$. This is done with the aid of an auxiliary polynomial of degree k that is *irreducible* (i.e., one that cannot be factorized using modulo p arithmetic).

I shall not go into details of Galois theory. My aim is simply to show, by means of a few examples, how F_q works when q is a power of a prime, so that finite projective geometries P(N, q) can be defined.

The four elements of $F_4 =$ GF(4) are

$$0, \quad 1, \quad x, \quad 1+x$$

in which the coefficients are integers modulo 2. They can be denoted by

$$(00), \quad (10), \quad (01), \quad (11).$$

Addition is obvious. For multiplication we need an auxiliary irreducible quadratic. All the quadratics are got by simply adding x^2 to each of the above four elements. The *reducible* quadratics are got by multiplying linear factors; there are three:

7.5 Galois Fields

x^2, $x(1+x) = x + x^2$ and $(1+x)(1+x) = 1 + x^2$, so in this case there is only *one* irreducible quadratic polynomial, $1 + x + x^2$. We *define*

$$1 + x + x^2 = 0.$$

Multiplication of a pair of elements of F_4 is done by multiplying in the ordinary way followed by reduction using this auxiliary equation to eliminate any x^2 that may have arisen. Modulo 2 arithmetic is used throughout. For example:

$$x \times x = x^2 = 1 + x$$
$$x \times (1 + x) = x + x^2 = x + (x + 1) = 1$$
$$(1 + x) \times (1 + x) = 1 + x^2 = 1 + (x + 1) = x$$

or, in the simper notation, $(01)(01) = (11)$, $(01)(11) = (10)$ and $(11)(11) = (01)$.

An alternative notation for the four elements of F_4 is

$$0, \quad 1, \quad \omega, \quad \bar\omega$$

The non-zero elements are analogous to the 'cube roots of unity' because, as is easily verified, they satisfy $\bar\omega = \omega^2$, $\omega\bar\omega = \omega^3 = 1$ and $\bar\omega + \omega + 1 = 0$. Interchange of ω and $\bar\omega$ is analogous to complex conjugation. More generally, raising all the elements of $GF(p^k)$ (p prime) to the pth power generates the *Frobenius automorphisms* on $GF(p^k)$ (this has no effect if $k = 1$ because of Fermat's 'little' theorem $a^p = a$ modulo p). The Frobenius automorphisms constitute a cyclic group of order k.

As a second example, let us look at GF(9). The elements are the nine polynomials

$$0, \quad \pm 1, \quad \pm x, \quad \pm 1, \quad \pm x$$

The arithmetic here, of course, is modulo 3. There are *three* irreducible quadratic equations. Now, a fundamental result in Galois field theory is that there is a *unique* field F_q for every q, so it *does not matter* which irreducible polynomial we use to define multiplication—the structures of the resulting F_qs are equivalent. For F_9, let us choose the irreducible quadratic

$$x^2 = -1$$

because it is the simplest to apply. We have, as an example of multiplication in F_9,

$$(1 + x)(1 + x) = 1 - x + x^2 = 1 - x + (-1) = -x.$$

That is, $(1\ 1)(1\ 1) = (0\ -1)$.

As a third and final example, consider F_8. The eight polynomials are

$$0, \quad 1, \quad x, \quad 1+x, \quad x^2, \quad 1+x^2, \quad x+x^2, \quad 1+x+x^2.$$

There are eight *cubic* polynomials, got by just adding an x^3 to these. Three of them are irreducible: $1+x^2+x^3$, $1+x+x^2+x^3$ and x^3+x+1. Any one of these three can be chosen to enable us to multiply the elements

$$(000), \quad (100), \quad (010), \quad (110), \quad (001), \quad (101), \quad (011), \quad (111)$$

of F_8.

7.6 PG(2, 4)

As a simple example of a projective geometry $PG(N, q)$, where q is not a prime but a power of a prime, consider the plane $PG(2, 4)$. The whole plane is a configuration (21_5). The homogeneous coordinates of points and lines are given by triples of elements belonging to F_4, namely the symbols 0, 1, ω and $\bar{\omega}$ that we just met, which can alternatively be written as (00), (10), (01) and (11) and represent the four expressions 0, 1, x and $1+x$. The arithmetic is modulo 2.

The four triangles whose vertices are given by the columns of the four matrices

$$\begin{pmatrix} 1 & 0 & 0 \\ 0 & 1 & 0 \\ 0 & 0 & 1 \end{pmatrix} \begin{pmatrix} 1 & 1 & 1 \\ 1 & \omega & \bar{\omega} \\ 1 & \bar{\omega} & \omega \end{pmatrix} \begin{pmatrix} \omega & 1 & 1 \\ 1 & \omega & 1 \\ 1 & 1 & \omega \end{pmatrix} \begin{pmatrix} \bar{\omega} & 1 & 1 \\ 1 & \bar{\omega} & 1 \\ 1 & 1 & \bar{\omega} \end{pmatrix}$$

have a remarkable property. *Any* two of them are perspective *in six different ways*! In each case the vertex and the axis of the perspectivity is a vertex and an edge belonging to the other pair of triangles. All this is easily verified by looking at the collinearities satisfied by sets of three of these 12 points.

The sides of these four triangles are given by the rows of the four matrices

$$\begin{bmatrix} 1 & 0 & 0 \\ 0 & 1 & 0 \\ 0 & 0 & 1 \end{bmatrix} \begin{bmatrix} 1 & 1 & 1 \\ 1 & \bar{\omega} & \omega \\ 1 & \omega & \bar{\omega} \end{bmatrix} \begin{bmatrix} \bar{\omega} & 1 & 1 \\ 1 & \bar{\omega} & 1 \\ 1 & 1 & \bar{\omega} \end{bmatrix} \begin{bmatrix} \omega & 1 & 1 \\ 1 & \omega & 1 \\ 1 & 1 & \omega \end{bmatrix}.$$

The intersections of the sides determine nine more points

$$\begin{array}{lll}
(0 \ 1 \ 1) & (1 \ 0 \ 1) & (1 \ 1 \ 0) \\
(0 \ 1 \ \omega) & (\omega \ 0 \ 1) & (1 \ \omega \ 0) \\
(0 \ 1 \ \bar{\omega}) & (\bar{\omega} \ 0 \ 1) & (1 \ \bar{\omega} \ 0).
\end{array}$$

We have met a set of nine points like this before, in connection with the inflections of a cubic curve in $P(2, C)$ (there are of course no minus signs this time because the arithmetic is modulo 2). They are collinear in threes on 12 lines—the 12 sides of the four triangles. They are the nine points of a *Jacobi configuration* $(9_4 12_3)$.

We can dualize all this. There are nine lines joining pairs of vertices

$$\begin{array}{lll}
[0 \ 1 \ 1] & [1 \ 0 \ 1] & [1 \ 1 \ 0] \\
[0 \ 1 \ \omega] & [\omega \ 0 \ 1] & [1 \ \omega \ 0] \\
[0 \ 1 \ \bar{\omega}] & [\bar{\omega} \ 0 \ 1] & [1 \ \bar{\omega} \ 0]
\end{array}$$

7.7 Structure of PG(N, q)

forming a dual Jacobi configuration (12_394). These two configurations (9_412_3) and (12_394) together account for all 21 points and 21 lines of PG(2, 4)—a (21_5).

7.7 Structure of PG(N, q)

Recall (page 54) how information about the incidence structures of the configurations α_N (simplexes) can be immediately extracted from Pascal's triangle of binomial coefficients

$$
\begin{array}{ccccccc}
 & & & 1 & & & \\
 & & 1 & & 1 & & \\
 & 1 & & 2 & & 1 & \\
1 & & 3 & & 3 & & 1 \\
\end{array}
$$
$$1 \quad 4 \quad 6 \quad 4 \quad 1$$
$$1 \quad 5 \quad 10 \quad 10 \quad 5 \quad 1$$
$$\cdots\cdots$$

For example α_4 is a configuration

5	4	6	4
2	10	3	3
3	3	10	2
4	6	4	5

The diagonal of the array lists the number of points, lines, planes and 3-spaces and the off-diagonal elements list the number of i-spaces incident with each j-space. The relation of the array to the triangle of binomial coefficients is immediately apparent, and generalizable in an obvious way for any value of N.

The purpose of this section is to extend this observation to the configuration PG(N, q).

For the binomial coefficients, $\binom{n}{0} = \binom{n}{n} = 1$ and the recurrence relation

$$\binom{n}{r} = \binom{n-1}{r-1} + \binom{n-1}{r}$$

allows Pascal's triangle to be rapidly generated.

Now, for any value of q, we introduce the generalized coefficients $\binom{n}{r}_q$ given by $\binom{n}{0}_q = \binom{n}{n}_q = 1$ and the recurrence relation

$$\binom{n}{r}_q = \binom{n-1}{r-1}_q + q^r \binom{n-1}{r}_q.$$

For example, for $q = 2$ we get the *generalized Pascal triangle*

$$
\begin{array}{ccccccc}
 & & & 1 & & & \\
 & & 1 & & 1 & & \\
 & & 1 & 3 & 1 & & \\
 & 1 & 7 & & 7 & 1 & \\
 1 & 15 & 35 & & 15 & 1 & \\
1 & 31 & 155 & 155 & 31 & 1 \\
 & & & \cdots\cdots & & &
\end{array}
$$

(e.g., $1+2\cdot 3 = 7$; $3+2^2\cdot 1 = 7$; $35+2^3\cdot 15 = 15+2^2\cdot 35 = 155$). From this triangle we can immediately produce the arrays describing, for example, the configurations PG(2, 2), PG(3, 2), PG(4, 2). They are

7	3
3	7

15	3	3
3	35	7
3	7	15

31	15	35	15
3	155	7	7
7	7	155	3
15	35	15	31

All this is surprising. It requires proof.

Let n_i be the number of i-spaces in PG(N, q). We have seen already that

$$n_0 = 1 + q + q^2 + \cdots + q^N. \tag{7.1}$$

Duality tells us that

$$n_i = n_{N-i-1}. \tag{7.2}$$

Suppose each i-space is incident with n_{ij} of the j-spaces. When $i < j$ this is the number of j-spaces that contain a given i-space and for $i > j$ it is the number of j-spaces that are contained in an i-space, so the duality tells us that

$$n_{ij} = n_{(N-i-1)(N-j-1)}. \tag{7.3}$$

The configuration of the i-spaces and j-spaces for some chosen i and j is an $(\alpha_\gamma \beta_\delta)$, with $\alpha = n_i$, $\beta = n_j$, $\gamma = n_{ij}$ and $\delta = n_{ji}$, so we have

$$n_i n_{ij} = n_j n_{ji}. \tag{7.4}$$

We now have enough information to determine all the n_i and the n_{ij}. From (7.4) we have

$$n_i n_{i(i-1)} = n_{i-1} n_{(i-1)i}. \tag{7.5}$$

From (7.1) and (7.2),

$$n_{i(i-1)} = n_{i0} = (q^{i+1} - 1)/(q - 1)$$

7.7 Structure of PG(N, q)

and from (7.1), (7.2) and (7.3),

$$n_{(i-1)i} = n_{(N-i)(N-i-1)} = n_{(N-1)0} = (q^{N-i+1} - 1)/(q - 1).$$

Substituting these values into (7.5) gives the recurrence relation

$$n_i = (q^{N-i-1} - 1)n_{i-1}/(q^{i+1} - 1). \tag{7.6}$$

Hence, for $j < i$ (since each i-space is a PG(i, q))

$$n_{ij} = (q^{i-j+1} - 1)n_{i(j-1)}/(q^{j+1} - 1). \tag{7.7}$$

(Observe that (7.7) includes (7.6) if we define $n_{Nj} = n_j$.) Also, since each i-space is a PG(i, q), Eq. (7.1) tells us that

$$n_{i0} = (q^{i+1} - 1)/(q - 1). \tag{7.8}$$

Equations (7.7) and (7.8) provide us with all the n_{ij} for which $j < i$. The result is

$$n_{ij} = \binom{i+1}{j+1}_q \quad (j < i). \tag{7.9}$$

where

$$\binom{n}{r}_q = \left(\frac{q^n - 1}{q - 1}\right)\left(\frac{q^{n-1} - 1}{q^2 - 1}\right) \cdots \left(\frac{q^{n-r+1} - 1}{q^{r-1} - 1}\right). \tag{7.10}$$

That the recurrence relation

$$\binom{n}{r}_q = \binom{n-1}{r-1}_q + q^r \binom{n-1}{r}_q$$

is satisfied is easily verified.

Dividing top and bottom of each factor of this expression by $q - 1$ we get

$$\binom{n}{r}_q = (q^{n-1} + q^{n-2} + \cdots + 1)\left(\frac{q^{n-2} + q^{n-3} + \cdots + 1}{q + 1}\right)$$

$$\times \left(\frac{q^{n-3} + q^{n-4} + \cdots + 1}{q^2 + q + 1}\right) \cdots \left(\frac{q^{n-r} + q^{n-r-1} + \cdots + 1}{q^{r-1} + q^{r-2} + \cdots + 1}\right)$$

and therefore

$$\binom{n}{r}_1 = n\left(\frac{n-1}{2}\right)\left(\frac{n-2}{3}\right) \cdots \left(\frac{n-r+1}{r}\right) = \binom{n}{r}.$$

From (7.3) and (7.7) we deduce that

$$n_{ij} = \binom{N-i}{j-i}_q \quad (j > i). \tag{7.11}$$

Equations (7.9) and (7.11) together completely determine the configuration of PG(N, q).

The surprising symmetry of the 'generalized Pascal triangle',

$$\binom{n}{r}_q = \binom{n}{n-r}_q$$

is easily verified directly from the definition, Eq. (7.10).

7.8 Collineations of PG(N, q)

In a projective N-space a unique collineation is determined if any $N + 2$ general points, of which no $N + 1$ of them lie in a hyperplane, and their respective $N + 2$ images are given. The first set of $N + 2$ can be chosen as the reference simplex and unit point. A unique collineation is then determined if the $N + 2$ images of the vertices of the reference simplex and the unit point are given. From this, it can be deduced that:

The collineations on PG(N, q) *constitute a group* LF(M, q), $M = N + 1$, *of order* $\Gamma(M, q) = q^{M(M-1)/2}(q^2 - 1)(q^3 - 1) \cdots (q^M - 1)/(M, q - 1)$.

The symbol (M, $q - 1$) denotes the greatest common factor of M and q.

Proof We need to find out the number of distinct ways of choosing $N + 2$ points in PG(N, q) (in a definite order) so that no $N + 1$ of them lie in an $(N - 1)$-space. The first point can be chosen in n_0 ways and the second in $n_0 - 1$ ways. The third point must not be collinear with these two. It can be chosen in $n_0 - n_{10}$ ways. The fourth must not be coplanar with these three, so can be chosen in $n_0 - n_{20}$ ways, and so on. We have, then,

$$n_0(n_0 - 1)(n_0 - n_{10})(n_0 - n_{20}) \cdots (n_0 - n_{(N-1)0})$$

ways of choosing $M = N + 1$ points. A simplex has then been selected. The final point to be chosen must not lie in any of its $(N - 1)$-dimensional faces. There are $(q - 1)^N$ points that do not lie on any face of a given simplex. (This can be seen by noting that the points that do not lie on any face of the *reference* simplex have no zero coordinates. There are $q - 1$ non-zero numbers in F_q, from which we can make $(q - 1)^M$ M-tuples, which are homogeneous coordinates for $(q - 1)^M/(q - 1) = (q - 1)^N$ distinct points.) There are therefore $(q - 1)^N$ choices for the final point. The number of ways of choosing an ordered set of $N + 2$ points, no $N + 1$ of which lie in an $(N - 1)$-space, is then

$$(q - 1)^N n_0(n_0 - 1)(n_0 - n_{10})(n_0 - n_{20}) \cdots (n_0 - n_{(N-1)0}).$$

Now substitute $n_0 = (q^{N+1} - 1)/(q - 1)$ and $n_{i0} = (q^{i+1} - 1)/(q - 1)$ and we get

$$(q - 1)^{-1}(q^M - 1)(q^M - q)(q^M - q^2) \cdots (q^M - q^N)$$

$$= (q-1)^{-1} q^{1+2+3+\cdots+N} (q^M - 1)(q^{M-1} - 1) \cdots (q^2 - 1)(q - 1)$$
$$= q^{M(M-1)/2} (q^M - 1)(q^{M-1} - 1) \cdots (q^2 - 1)$$

ways of choosing the (N + 2) points. There is a unique collineation that will convert the reference points and unit points to the points of the chosen set. Its matrix can, without loss of generality, be made unimodular by dividing out its determinant. We now have a matrix of the *special linear group* SL(M, q). But it still may not be the *unique* matrix for the collineation because, unlike the real and complex cases, SL(N, q) may contain multiples of the unit matrix—multiples κ of the unit matrix for which $\kappa^M = 1$. The number of elements of κ of F_q for which $\kappa^{N+1} = 1$ is (M, $q-1$) (I will not present a proof—it is a consequence of 'Fermat's little theorem' that $a^{p-1} = 1$ modulo p). These have to be factored out. The resulting group of collineations LF(M, q) is therefore a factor group SL(M, q)/K where SL(M, q) is the group of M × M unimodular matrices whose elements are members of F_q and K is the group of multiples of the unit matrix with unit determinant.

The result follows. The *linear fractional group* LF(M, q) of collineations of PG(N, q) is a group of order Γ(M, q).

LF(M, q) is a subgroup of the full *symmetry* group of the configuration formed by all the points, lines, planes, etc.—not all symmetries are necessarily realizable as collineations. If $q = p^k$ the complete symmetry is generated by the collineations together with the Frobenius automorphisms generated by $a \to a^p$ applied to all the elements a of F_q—a cyclic group of order k. These are symmetries of the configuration that cannot be brought about by collineations.

7.9 Finite Projective Lines

The one-dimensional geometry PG(1, q) is the geometry of the *projective line* PL(q). It consists simply of $q + 1$ 'points', labeled by a *homographic parameter* θ. The $q + 1$ values of θ are the q elements of F_q and an extra element denoted by ∞. The homographies are the transformations

$$\theta \to \frac{a\theta + b}{c\theta + d}, \quad ad - bc = 1$$

(including $\infty \to a/c$). All that we said about homographies on the real projective line or the complex projective line applies to these finite lines, except that now the elements a, b, c and d of the unimodular matrices

$$\begin{pmatrix} a & b \\ c & d \end{pmatrix}$$

are elements of the finite field F_q. They belong to the *special linear group* SL(2, q). However, multiples of the unit matrix may have unit determinant if F_q has elements a satisfying $a^2 = 1$ (e.g. $2^2 = 1$ modulo 3). Eliminating these by forming a factor group we have the *linear fractional group* LF(2, q), which we shall henceforth abbreviate to LF_q. It permutes the $q+1$ 'points' and so is a subgroup of the group S_{q+1} of permutations of $q+1$ objects.

LF_q is generated by the three homographies

$$\alpha: \quad \theta \to \theta+1 \qquad \beta: \quad \theta \to k\theta \qquad \gamma: \quad \theta \to -1/\theta$$

provided k is chosen so that the square of every non-zero element of F_q is a power of k. As is not difficult to verify, $\theta \to (\theta+b)/(c\theta+d)$, $ad - bc = 1$, is given by $\beta^\mu \alpha^{b/a}$ ($k^\mu = a^2$) if $c = 0$ and by $\beta^\mu \alpha^{ac} \gamma \alpha^{d/c}$ ($k^\mu = c^{-2}$) if $c \neq 0$.

The linear fractional groups play a fundamental role in the intricate theory of finite groups, which is why the finite projective lines are of interest. Besides which, these groups have interesting characteristics—each of them has a unique personality, as we shall see.

In the following sections we shall look briefly at some of the permutation groups LF_p (p a prime). Much of this material is selected and abridged from John Conway's remarkable 'Three Lectures on Exceptional Groups'.

7.10 PL(3)

The homographies

$$\theta \to \frac{a\theta + b}{c\theta + d}, \quad ad - bc = 1$$

of PL(3) permute the four values 0 1 2 ∞ of the parameter θ. The whole group LF_3 is generated by the two homographies

$$\alpha: \quad \theta \to \theta+1 \qquad \gamma: \quad \theta \to -1/\theta$$

(modulo 3 arithmetic). These correspond to the permutations

$$\alpha = (012) \qquad \gamma = (0\infty)(12).$$

But these permutations generate the group A_4 of all the even permutations of 0 1 2 ∞ (a group of order 12), so we have an isomorphism $LF_3 \sim A_4$.

7.11 PL(5)

The homographies of PL(5) permute the six values 0 1 2 3 4 ∞ of the parameter θ. That is, LF_5 is a subgroup of the group S_6 of permutations of six objects. It is generated by the three homographies

$$\alpha: \quad \theta \to \theta+1 \qquad \beta: \quad \theta \to -\theta \qquad \gamma: \quad \theta \to -1/\theta$$

7.11 PL(5)

(modulo 5 arithmetic) which produce the permutations

$$\alpha = (01234) \quad \beta = (14)(23) \quad \gamma = (0\infty)(14).$$

The whole of S_6 is generated by α and $\tau = (0\infty)$ or by α and $\pi = (0\infty)(14)(23)$. We have already remarked that S_6 possesses outer automorphisms. For example, the automorphism ϕ given by

$$\alpha \to \alpha \quad \tau \to \pi \quad \pi \to \tau$$

interchanges transpositions and products of three transpositions.

The action of the powers of α on τ gives the five transpositions

$$\tau_0 = (0\infty) \quad \tau_1 = (1\infty) \quad \tau_2 = (2\infty) \quad \tau_3 = (3\infty) \quad \tau_4 = (4\infty)$$

and the action of powers of α on π gives five products of three transpositions

$$\pi_0 = (0\infty)(14)(23)$$
$$\pi_1 = (1\infty)(20)(34)$$
$$\pi_2 = (2\infty)(31)(40)$$
$$\pi_3 = (3\infty)(42)(01)$$
$$\pi_4 = (4\infty)(03)(12).$$

(The action of a permutation μ on a permutation ν is $\mu\nu\mu^{-1}$ which may be conveniently written as ν^μ.) Now the action of the subgroup LF_5 of S_6 permutes $\tau_0\tau_1\tau_2\tau_3\tau_4$. It also permutes $\pi_0\pi_1\pi_2\pi_3\pi_4$. Specifically,

$$\tau_\theta{}^\alpha = \tau_{\alpha(\theta)} \quad \tau_\theta{}^\beta = \tau_{\beta(\theta)} \quad \tau_\theta{}^\gamma = \tau_{\delta(\theta)}$$
$$\pi_\theta{}^\alpha = \pi_{\alpha(\theta)} \quad \pi_\theta{}^\beta = \pi_{\beta(\theta)} \quad \pi_\theta{}^\gamma = \pi_{\delta(\theta)}$$

where

$$\delta = (12)(34).$$

The surprise here is that LF_5, which was defined as a group of permutations of *six* objects, is now represented by permutations that act on only *five* objects. We have an LF_5 generated by

$$\alpha = (01234) \quad \beta = (14)(23) \quad \delta = (12)(34).$$

The automorphism ϕ of S_6 that interchanges the two kinds of LF_5 subgroup is

$$\alpha \to \alpha \quad \beta \to \beta \quad \gamma \leftrightarrow \delta.$$

Note that α, β and δ are all *even* permutations. The group generated by α, β and δ is in fact the *alternating group* A_5 of all even permutations of five objects. There is therefore an *isomorphism* $LF_5 \sim A_5$.

Fig. 7.2 Six five-fold axes of an icosahedron

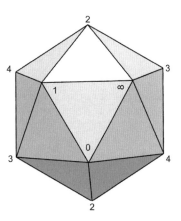

Recall how the outer automorphisms of S_6 were introduced by way of Sylvester's duads and synthemes. In the present context the 15 transpositions in S_6 are the *duads* $\tau_{ij} = \tau_j{}^{\tau_i} = \tau_i{}^{\tau_j}$ ($i, j = 0, \ldots, 4$) and $\tau_{i\infty} = \tau_i$. Similarly, the 15 products of three transpositions are *synthemes* $\pi_{ij} = \pi_j{}^{\pi_i} = \pi_i{}^{\pi_j}$ ($i, j = 0, \ldots, 4$) and $\pi_{i\infty} = \pi_i$. The following table lists all 15 synthemes (e.g. $\pi_{13} = (30)(14)(2\infty)$) and also displays the action of ϕ on the duads (e.g. ϕ maps (13) to $(30)(14)(2\infty)$).

	0	1	2	3	4	∞
0	–	(12)(3∞)(04)	(24)(1∞)(03)	(31)(02)(4∞)	(43)(2∞)(01)	(∞0)(23)(14)
1	(12)(3∞)(04)	–	(23)(4∞)(01)	(30)(14)(2∞)	(42)(13)(0∞)	(∞1)(34)(20)
2	(24)(1∞)(03)	(23)(4∞)(01)	–	(34)(0∞)(12)	(41)(20)(∞3)	(∞2)(13)(04)
3	(31)(02)(4∞)	(30)(14)(2∞)	(34)(0∞)(12)	–	(40)(1∞)(23)	(∞3)(24)(01)
4	(43)(2∞)(01)	(42)(13)(0∞)	(41)(20)(∞3)	(40)(1∞)(23)	–	(∞4)(03)(12)
∞	(∞0)(23)(14)	(∞1)(34)(20)	(∞2)(13)(04)	(∞3)(24)(01)	(∞4)(03)(12)	–

(Notice, incidentally, how this table is related to the table of Sylvester's synthemes—if we simply replace the symbols 01234∞, respectively, by **123456**. Indeed, that is how I constructed that particular table of synthemes.)

The fact that A_5 can be represented as a group of permutations of *six* objects has an interesting interpretation in terms of a familiar figure in Euclidean 3-space—the *regular icosahedron*. The group of rotational symmetries of a regular icosahedron is A_5—a group of order 60. The icosahedron has six five-fold axes, labeled in Fig. 7.2. The rotations are generated by the permutations α, β and γ applied to these axes. Five objects permuted by this action can be taken to be the five trios of mutually perpendicular 'golden' rectangles (Fig. 7.3). A golden rectangle can be denoted by a pair of symbols indicating the two axes that are its diagonals. The five objects are then

$$\pi_0 = (0\infty)(14)(23)$$
$$\pi_1 = (1\infty)(20)(34)$$
$$\pi_2 = (2\infty)(31)(40)$$

7.11 PL(5)

Fig. 7.3 A trio of golden rectangles

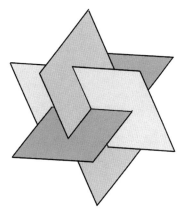

Fig. 7.4 Second kind of trio of rectangles

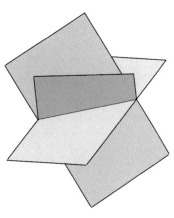

Fig. 7.5 Compound of five tetrahedra in an icosahedron

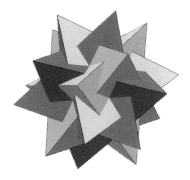

$$\pi_3 = (3\infty)(42)(01)$$
$$\pi_4 = (4\infty)(03)(12)$$

The remaining ten *synthemes* correspond to 'trios of the second kind' (Fig. 7.4). The 15 *duads* correspond, in an obvious way, to the 15 rectangles or, equivalently, the

15 two-fold axes of the icosahedron. Another five objects permuted by α, β and δ when the six five-fold axes are permuted by α, β and γ are five regular tetrahedra (Fig. 7.5) whose vertices are mid-points of the faces of the icosahedron.

7.12 Six points in PG(2, 4)

The columns of

$$\begin{pmatrix} 1 & 0 & 0 & 1 & \omega & \omega \\ 0 & 1 & 0 & \omega & 1 & \omega \\ 0 & 0 & 1 & \omega & \omega & 1 \end{pmatrix}$$

are homogeneous coordinates for six points in PG(2, 4). Call these six points **14∞230** (in that order). Then **14∞** and **230** are two triangles which, as we have already noted, are perspective in six different ways. The collineations with matrices

$$A = \begin{pmatrix} 1 & \omega & 0 \\ \omega & \omega & 0 \\ \omega & 1 & 1 \end{pmatrix} \quad B = \begin{pmatrix} 0 & 1 & 0 \\ 1 & 0 & 0 \\ 0 & 0 & 1 \end{pmatrix}$$

$$C = \begin{pmatrix} 0 & 1 & \omega \\ 1 & 0 & \omega \\ 0 & 0 & 1 \end{pmatrix} \quad D = \begin{pmatrix} 1 & \omega & 0 \\ \omega & 1 & 0 \\ \omega & \omega & 1 \end{pmatrix}$$

permute the six points according to the permutations α, β, γ and δ of LF$_5$. (Remember when verifying this that the arithmetic is that of the Galois field F$_4$!) So there are two different kinds of LF$_5$ groups of collineations of PG(2, 4) that leave the six points invariant—one generated by A, B, and C and one generated by A, B and D. But A, B and D leave the line [0 0 1] ($x^3 = 0$) invariant and therefore this LF$_5$ acts as a group of homographies on this line. But this line is a PL$_4$ whose homography group is LF$_4$. So LF$_5$ is a subgroup of LF$_4$. But LF$_5$ has order $5(5^2 - 1)/2 = 60$ and LF$_4$ has order $4(4^2 - 1) = 60$. Hence the rather surprising isomorphism

$$\text{LF}_5 \sim \text{LF}_4.$$

7.13 PL(7)

The group LF$_7$ acts on the eight values 0 1 2 3 4 5 6 ∞ of the parameter θ. It is generated by the three homographies

$$\alpha: \quad \theta \to \theta + 1 \qquad \beta: \quad \theta \to 2\theta \qquad \gamma: \quad \theta \to -1/\theta$$

(modulo 7 arithmetic), which produce the permutations

$$\alpha = (0123456) \qquad \beta = (124)(365) \qquad \gamma = (0\infty)(16)(23)(45).$$

Fig. 7.6 Fano's configuration

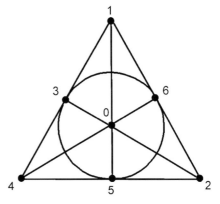

Starting with $\pi = (\infty 0)(15)(23)(46)$ and proceeding as we did for PL(5), we find there is a group isomorphic to LF$_7$ that acts on only *seven* objects, generated by α, β and δ, where

$$\delta = (14)(56).$$

In Fig. 7.6 the vertices of Fano's (7_3) are labeled with the seven values of θ on which this second LF$_7$ acts. With this labeling we see that α cyclically permutes the vertices of a 'self-inscribed and circumscribed heptagon', that β rotates the figure, and δ reflects it. The permutations α, β and δ generate the group of collineations of PG(2, 2). We have another surprising isomorphism

$$\text{LF}_7 \sim \text{LF}(3, 2).$$

This is the symmetry group of the Fano configuration, which, as we already deduced by counting quadrilaterals, is a group of order 168.

7.14 Eight Points in PG(3, 2)

The columns of the matrix

$$\begin{pmatrix} 1 & 0 & 0 & 0 & 0 & 1 & 1 & 1 \\ 0 & 1 & 0 & 0 & 1 & 0 & 1 & 1 \\ 0 & 0 & 1 & 0 & 1 & 1 & 0 & 1 \\ 0 & 0 & 0 & 1 & 1 & 1 & 1 & 0 \end{pmatrix}$$

are homogeneous coordinates for eight points in PG(3, 2). Call the eight points/columns **124∞** and **5360** (in that order). As is easily verified, the two *tetrahedra* **124∞** and **5360** are *mutually inscribed*—each vertex of one of them is contained in a face of the other (the arithmetic for verifying this is, of course, modulo 2). A pair of tetrahedra with this property is a *Möbius configuration*, an (8_4) of points and planes.

Then the permutations α, β, γ and δ of LF$_7$ applied to this set of eight points are brought about by the *collineations* of PG(3, 2) whose matrices are

$$A = \begin{pmatrix} 0 & 1 & 0 & 0 \\ 1 & 0 & 1 & 0 \\ 0 & 1 & 1 & 0 \\ 0 & 1 & 1 & 1 \end{pmatrix} \quad B = \begin{pmatrix} 0 & 0 & 1 & 0 \\ 1 & 0 & 0 & 0 \\ 0 & 1 & 0 & 0 \\ 0 & 0 & 0 & 1 \end{pmatrix}$$

$$C = \begin{pmatrix} 1 & 1 & 0 & 1 \\ 1 & 0 & 1 & 1 \\ 0 & 1 & 1 & 1 \\ 1 & 1 & 1 & 0 \end{pmatrix} \quad D = \begin{pmatrix} 0 & 0 & 1 & 0 \\ 0 & 1 & 0 & 0 \\ 1 & 0 & 0 & 0 \\ 0 & 0 & 0 & 1 \end{pmatrix}$$

—as is easily seen by multiplying the matrices that represent the two tetrahedra by these matrices. PG(3, 2) contains in all $1 + 2 + 2^2 + 2^3 = 15$ points. The eight points of the (8_4) configuration are all those with an odd number of 1s in their set of coordinates. The remaining seven points *0 1 2 3 4 5 6* are all those with an even number of 1s—their coordinates are given by the seven columns of

$$\begin{pmatrix} 1 & 1 & 0 & 1 & 0 & 0 & 1 \\ 1 & 0 & 1 & 0 & 0 & 1 & 1 \\ 1 & 0 & 0 & 1 & 1 & 1 & 0 \\ 1 & 1 & 1 & 0 & 1 & 0 & 0 \end{pmatrix}.$$

They are all the points that lie in the plane [1111] and are, therefore, the points of a Fano configuration PG(2, 2). The unit plane can be chosen as the *plane at infinity* and then we see that the (8_4) is the entire *affine* 3-space over the finite field F_2. The collineations A, B, and D produce the permutations α, β and δ of these seven points at infinity.

The subgroup G of FL(4, 2) is generated by α, β, γ and δ. But $\gamma\delta = \pi(!)$ so this group is also, alternatively, generated by α, β, δ and π. The eight generators $\pi_0, \pi_1, \ldots, \pi_7$ commute, so G has an invariant Abelian subgroup of order 8. It is a group of order $8 \cdot 168 = 1344$. As we have seen, it permutes the eight points in PG(3, 2) that we began with. Looking again at the coordinates of these eight points, we see that **1240** are coplanar, and so are **536∞**. (See the pattern: 1240 are the quadratic residues—perfect squares—in F_7.) By repeated action of the collineation A we get 14 *planes*

Q_0	0124	N_0	536∞
Q_1	1235	N_1	640∞
Q_2	2346	N_2	051∞
Q_3	3450	N_3	162∞
Q_4	4561	N_4	203∞
Q_5	5602	N_5	314∞
Q_6	6123	N_6	425∞

each containing four of the points, and every set of three of the points lies in just one of these planes. Hence the Möbius (8_4) is a *Steiner system* S(3, 4, 8)—which has $\binom{8}{3}/\binom{4}{3} = 14$ 'tetrads'. The collineation group G is the symmetry group of this Steiner system.

7.15 Steiner Systems

A *Steiner system* S(l, m, n) is a set of n 'things' containing $\binom{n}{l}/\binom{m}{l}$ sets of m—called *blocks* in general and *m-ads* in particular cases—so that any set of l of the things belongs to just one *block*. For example, a finite plane PG(2, q) is a Steiner system S(2, $q+1$, q^2+q+1), since it has q^2+q+1 points, with each line passing through $q+1$ points and each pair of distinct points lying on exactly one line.

There are two Steiner systems of very special interest, which we shall soon meet. S(5, 6, 12) is a set of 12 things containing $\binom{12}{5}/\binom{6}{5} = 132$ sets of 6 (the *hexads*) so that any 5 of the things belongs to just one hexad. S(5, 8, 24) is a set of 24 things containing $\binom{24}{5}/\binom{8}{5} = 759$ sets of 8 (the *octads*) so that any 5 of the things belongs to just one octad.

7.16 PL(11)

LF$_{11}$ acts on the 12 points 0 1 2 3 4 5 6 7 8 9 X ∞ of PL(11) and is generated by

$$\alpha: \quad \theta \to \theta+1 \qquad \beta: \quad \theta \to 3\theta \qquad \gamma: \quad \theta \to -1/\theta$$

(modulo 11), which produce the permutations

$$\alpha = (0123456789X) \qquad \beta = (13954)(8267X)$$

$$\gamma = (0\infty)(1X)(25)(37)(48)(69).$$

Define $\pi = (0\infty)(16)(37)(9X)(58)(42)$ and by the now familiar procedure we find an LF$_{11}$ that acts on only 11 objects, generated by α, β and

$$\delta = (2X)(34)(59)(67).$$

Both kinds of LF$_{11}$ are subgroups of a larger group of permutations of the 12 points, generated by α, β, γ and δ. This is the *Mathieu group* M$_{12}$, a group of order 95040. It is the symmetry group of the *Steiner system* S(5, 6, 12). Eleven of these hexads will be seen in the next section.

7.17 Coxeter's Configuration (11$_6$)

An icosahedron has 12 vertices. A *hemi-icosahedron* is an icosahedron with diametrically opposite points identified. It has six vertices, ten faces and 15 edges. In

Fig. 7.7 A hemi-icosahedron

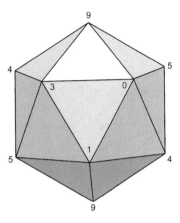

Fig. 7.7 the six vertices of a hemi-icosahedron have been labeled **0 1 3 9 5 4**. (These symbols are actually the 'quadratic residues' of F_{11}—the squares of integers modulo 11.) We shall give the name 0 to the hemi-icosahedron with this set of vertex labels.

Now produce a set of 11 hemi-icosahedra by repeatedly applying the permutation α of FL_{11} to the names of the hemi-icosahedra, and α^{-1} to the names of the vertices:

0	**013954**
1	**X02843**
2	**9X1732**
3	**890621**
4	**78X510**
5	**67940X**
6	**5683X9**
7	**457298**
8	**346187**
9	**235076**
X	**124X65**

This is a set of 11 objects and 11 vertices, with each object containing six vertices and with six of the objects sharing each vertex—a configuration (11_6) of hemi-icosahedra and vertices. Its self-duality is obvious from the fact that the table got from this one by simply interchanging bold and italic reveals all the same incidences.

The symmetry group of Coxeter's (11_6) is the fractional linear group LF_{11}.

The symmetries are generated by the action of α, β and δ on the vertices which, respectively, produce the corresponding actions of α^{-1}, β and δ on the hemi-icosahedra. The self-duality corresponds to the automorphism $\alpha \to \alpha^{-1}, \beta \to \beta, \delta \to \delta$ of F_{11}.

7.18 Twelve Points in PG(5, 3)   165

The subgroup FL_5 of FL_{11} that keeps one hemi-icosahedron invariant can be understood in terms of the correspondence

$$PL_5:\ 0\ \ 1\ \ 2\ \ 3\ \ 4\ \ \infty$$
$$Q:\ \ \ \ \ 1\ \ 3\ \ 9\ \ 5\ \ 4\ \ 0$$

between the points of PL_5 and the quadratic residues Q of F_{11}. (This is the correspondence between the labellings of Figs. 7.2 and 7.7.) Then we find that the β, δ and $\alpha^{-3}\delta\alpha^3$ of FL_{11} correspond to the α, β and γ of FL_5, respectively.

Notice how the hemi-icosahedra all 'fit together'—every triangular face is shared by just two of them (**395** for example belongs to 0 and 6). This is, of course, impossible to imagine! On the other hand, it is not difficult to deduce that the whole structure is a configuration

$$\begin{array}{|cccc|}\hline 11 & 10 & 15 & 6 \\ 2 & 55 & 3 & 3 \\ 3 & 3 & 55 & 2 \\ 6 & 15 & 10 & 11 \\ \hline\end{array}$$

7.18 Twelve Points in PG(5, 3)

The columns of

$$\begin{pmatrix} 1 & 0 & 0 & 0 & 0 & 0 & 0 & -1 & 1 & 1 & -1 & -1 \\ 0 & 1 & 0 & 0 & 0 & 0 & -1 & 0 & -1 & 1 & 1 & -1 \\ 0 & 0 & 1 & 0 & 0 & 0 & 1 & -1 & 0 & -1 & 1 & -1 \\ 0 & 0 & 0 & 1 & 0 & 0 & 1 & 1 & -1 & 0 & -1 & -1 \\ 0 & 0 & 0 & 0 & 1 & 0 & -1 & 1 & 1 & -1 & 0 & -1 \\ 0 & 0 & 0 & 0 & 0 & 1 & 1 & 1 & 1 & 1 & 1 & 0 \end{pmatrix}$$

are coordinates of 12 points in PG(5, 3). Denote the 12 vertices/columns by the values of the homographic parameter of PL(11):

1 3 9 5 4 ∞ 8 2 6 7 X 0

in that order. (Notice, incidentally, that the parameters 1 3 9 5 4 are perfect squares in F_{11} and 8 2 6 7 are these multiplied by 8.)

The two simplexes **13954∞** and **8267X0** are a Möbius pair—every vertex of each of them is contained in a (4-dimensional) face of the other; they are 'mutually inscribed'. For example, **1 3 9 5 4** and **0** all lie in the 4-space [000001] ($x^6 = 0$) and **8 2 6 7 X** and ∞ all lie in [111110]. (So we could have alternatively introduced the 12 points as a *pair of hexastigms* in two four-dimensional subspaces of PG(5, 3).)

The four collineations of PG(5, 3) given by

$$A = \begin{pmatrix} -1 & 0 & -1 & -1 & 0 & 0 \\ 0 & 0 & 1 & 1 & 0 & 0 \\ -1 & 0 & 1 & 1 & 0 & 0 \\ 1 & 0 & -1 & -1 & 1 & 0 \\ 1 & 1 & 0 & 0 & 0 & 0 \\ 1 & 0 & 1 & 1 & 0 & 1 \end{pmatrix} \quad B = \begin{pmatrix} 0 & 0 & 0 & 0 & 1 & 0 \\ 1 & 0 & 0 & 0 & 0 & 0 \\ 0 & 1 & 0 & 0 & 0 & 0 \\ 0 & 0 & 1 & 0 & 0 & 0 \\ 0 & 0 & 0 & 0 & 0 & 0 \\ 0 & 0 & 0 & 1 & 0 & 1 \end{pmatrix}$$

$$C = \begin{pmatrix} -1 & 1 & 1 & -1 & 0 & 1 \\ 1 & 1 & -1 & 0 & -1 & 1 \\ 1 & -1 & 0 & -1 & 1 & 1 \\ -1 & 0 & -1 & 1 & 1 & 1 \\ 0 & -1 & 1 & 1 & -1 & 1 \\ 1 & 1 & 1 & 1 & 1 & 0 \end{pmatrix} \quad D = \begin{pmatrix} 1 & 0 & 0 & 0 & 0 & 0 \\ 0 & 0 & 0 & 0 & 1 & 0 \\ 0 & 0 & 0 & 1 & 0 & 0 \\ 0 & 0 & 1 & 0 & 0 & 0 \\ 0 & 1 & 0 & 0 & 0 & 0 \\ 0 & 0 & 0 & 0 & 0 & 1 \end{pmatrix}$$

permute the 12 points according to the permutations α, β, γ and δ of FL_{11}.

The repeated action of the collineation A produces, from the two 4-spaces **139540** and **8267X∞**, a set of 22 4-spaces (11 pairs of complementary hexastigms)

Q_0	139540	N_0	8267X∞
Q_1	24X651	N_1	93780∞
Q_2	350762	N_2	X4891∞
⋮	⋮	⋮	⋮
Q_9	X17329	N_9	60458∞
Q_X	02843X	N_X	71569∞

There are altogether 132 hexastigms obtained by the permutations α, β, γ and δ of the 12 points. Twelve points in $\binom{12}{6}/\binom{6}{5} = 132$ hexastigms, every set of five of the points belongs to a unique hexastigm—we have here a Steiner system S(5, 6, 12). The symmetry group of S(5, 8, 12) is the *Mathieu group* M_{12}, of order 95040. In this context, it is generated by the permutations α, β, γ and δ and is realizable as a group of projective transformations in PG(5, 3).

7.19 PL(23)

LF_{23} acts on the 24 points

$$0\ 1\ 2\ 3\ 4\ 5\ 6\ 7\ 8\ 9\ 10\ 11\ 12\ 13\ 14\ 15\ 16\ 17\ 18\ 19\ 20\ 21\ 22\ 23\ \infty$$

of the projective line PL(23) and is generated by

$$\alpha: \ \theta \to \theta + 1 \qquad \beta: \ \theta \to 2\theta \qquad \gamma: \ \theta \to -1/\theta$$

(modulo 23) which produce the permutations

$\alpha = (0\ 1\ 2\ 3\ 4\ 5\ 6\ 7\ 8\ 9\ 10\ 11\ 12\ 13\ 14\ 15\ 16\ 17\ 18\ 19\ 20\ 21\ 22\ 23)$

$\beta = (1\ 2\ 4\ 8\ 16\ 9\ 18\ 13\ 3\ 6\ 12)(5\ 10\ 20\ 17\ 11\ 22\ 21\ 19\ 15\ 7\ 14)$

$\gamma = (0\ \infty)(1\ 22)(2\ 11)(4\ 17)(8\ 20)(16\ 10)(9\ 5)(18\ 14)(13\ 7)(3\ 15)(6\ 19)(12\ 21).$

We might expect, from what has gone before, that there might be another FL_{23} that acts on only 23 of the points, generated by α, β and δ, but this is not so. There is, however, an analogy with the fact that FL_{11} is a subgroup of M_{12}.

The Mathieu group M_{24} (of order 244823040) is the symmetry group of the Steiner system $S(5, 8, 24)$: a set of 24 things containing $\binom{24}{5}/\binom{8}{5} = 759$ sets of 8 (the *octads*) so that any 5 of the things belong to just one octad. M_{24} is generated by α, β, γ and

$$\delta = (14\ 17\ 11\ 19\ 22)(20\ 10\ 7\ 5\ 21)(18\ 4\ 2\ 6\ 1)(8\ 16\ 13\ 9\ 12).$$

It turns out that $\sigma = \gamma\delta$ satisfies $\sigma^5 = \gamma$ and $\sigma^6 = \delta$, so M_{24} is generated by α, β and σ.

7.20 Twenty-Four Points in PG(11, 2)

The columns of

$$\begin{pmatrix} 1 & & & & & & & & & & & & 1 & 1 & 0 & 1 & 1 & 1 & 0 & 0 & 0 & 1 & 0 & 1 \\ & 1 & & & & & & & & & & & 0 & 1 & 1 & 0 & 1 & 1 & 1 & 0 & 0 & 0 & 1 & 1 \\ & & 1 & & & & & & & & & & 1 & 0 & 1 & 1 & 0 & 1 & 1 & 1 & 0 & 0 & 0 & 1 \\ & & & 1 & & & & & & & & & 0 & 1 & 0 & 1 & 1 & 0 & 1 & 1 & 1 & 0 & 0 & 1 \\ & & & & 1 & & & & & & & & 0 & 0 & 1 & 0 & 1 & 1 & 0 & 1 & 1 & 1 & 0 & 1 \\ & & & & & 1 & & & & & & & 0 & 0 & 0 & 1 & 0 & 1 & 1 & 0 & 1 & 1 & 1 & 1 \\ & & & & & & 1 & & & & & & 1 & 0 & 0 & 0 & 1 & 0 & 1 & 1 & 0 & 1 & 1 & 1 \\ & & & & & & & 1 & & & & & 1 & 1 & 0 & 0 & 0 & 1 & 0 & 1 & 1 & 0 & 1 & 1 \\ & & & & & & & & 1 & & & & 1 & 1 & 1 & 0 & 0 & 0 & 1 & 0 & 1 & 1 & 0 & 1 \\ & & & & & & & & & 1 & & & 0 & 1 & 1 & 1 & 0 & 0 & 0 & 1 & 0 & 1 & 1 & 1 \\ & & & & & & & & & & 1 & & 1 & 0 & 1 & 1 & 1 & 0 & 0 & 0 & 1 & 0 & 1 & 1 \\ & & & & & & & & & & & 1 & 1 & 1 & 1 & 1 & 1 & 1 & 1 & 1 & 1 & 1 & 1 & 0 \end{pmatrix}$$

are the coordinates of 24 points in PG(11, 2).

This is how the matrix has been constructed: squaring the elements 0123456789X of F_{11}, we get 01495335941 (in that order), so the set of quadratic residues is 014953. Now replace each element by 1 if it is a quadratic residue and by 0 if it is not. We get 11011100010. See how the 11×11 matrix obtained by cyclically permuting this row occurs in the 24×12 matrix. This all looks very mysterious—a proper explanation would take us into the theory of error-correcting codes and this is supposed to be a *geometry* book! If your curiosity is aroused you will need to go to the relevant literature (see the Appendix for references). Suffice it to say that all the (modulo 2) linear combinations of the *rows* of our 24×12 matrix are the $2^{12} = 4096$ words of the *extended binary Golay code*. Similarly, the rows of the 6×3 matrix for the 'six points in PG(2, 4)' generate the *hexacode*, the rows of the 8 × 4 matrix for the eight points in PG(3, 2) generate the *Hamming code,* and the rows of the 12×6 matrix for the 12 points in PG(5, 3) generate the $3^6 = 729$ words of the *extended ternary Golay code*.

Label the 24 points/columns

$$1\ 2\ 4\ 8\ 16\ 9\ 18\ 13\ 3\ 6\ 12\ \infty \qquad 5\ 10\ 20\ 17\ 11\ 22\ 21\ 19\ 15\ 7\ 14\ 0$$

—in that order. We have here a pair of mutually inscribed simplexes. Alternatively, interchanging **0** and ∞ the configuration can be described as two complementary sets of 12 points, each set lying in a 10-space. Let us call a set of 12 general points in 10-space a *dodecastigm* (in analogy with hexastigm—six general points in a 4-space). The 24 points belong to a complementary pair of dodecastigms

$$Q: \quad 1\ 2\ 4\ 8\ 16\ 9\ 18\ 13\ 3\ 6\ 12\ \infty$$

$$N: \quad 5\ 10\ 20\ 17\ 11\ 22\ 21\ 19\ 15\ 7\ 14\ 0$$

We call them Q and N because 1 2 3 8 16 9 18 13 3 6 12 are the non-zero Quadratic residues (perfect squares) in F_{23}, which also happen to be the powers of 2. Multiplying by 5 we get N, which are Non-residues.

The permutations α, β and σ of PL(23) have corresponding *collineations* in PG(11, 2). I will not write them down; I will let you do that. The Mathieu group M_{24} is then realized as a group of collineations on PG(11, 2) that preserves all the incidences of the configuration determined by the 24 points.

The eight points **1 5 10 17 11 22 7 0** are all the points whose first coordinate is 1 (see the first row of the 24 × 12 matrix). Their coordinates are the columns of

$$\begin{pmatrix} 1 & 1 & 1 & 1 & 1 & 1 & 1 & 1 \\ 0 & 0 & 1 & 0 & 1 & 1 & 0 & 1 \\ 0 & 1 & 0 & 1 & 0 & 1 & 0 & 1 \\ 0 & 0 & 1 & 1 & 1 & 0 & 0 & 1 \\ 0 & 0 & 0 & 0 & 1 & 1 & 1 & 1 \\ 0 & 0 & 0 & 1 & 0 & 1 & 1 & 1 \\ 0 & 1 & 0 & 0 & 1 & 0 & 1 & 1 \\ 0 & 1 & 1 & 0 & 0 & 1 & 0 & 1 \\ 0 & 1 & 1 & 0 & 0 & 0 & 1 & 1 \\ 0 & 0 & 1 & 1 & 0 & 0 & 1 & 1 \\ 0 & 1 & 0 & 1 & 1 & 0 & 0 & 1 \\ 0 & 1 & 1 & 1 & 1 & 1 & 1 & 0 \end{pmatrix}$$

which are linearly dependent—they sum to zero, modulo 2. Symbolically,

$$1 + 5 + 10 + 17 + 11 + 22 + 7 + 0 = 0.$$

So these eight points all lie in a 6-space. I shall call a set of eight general points in a 6-space an *octastigm*. The question now is: how many of these octastigms can be formed from the 24 points? And how many dodecastigms?

First note that the final coordinate of every point in Q is 0 and that the final coordinate of every point in N is 1. Hence n of the points can all lie in an $(n-2)$-space only if an even number of them is in N.

There are $\binom{12}{2} = 66$ ways of choosing a pair of points from N. In each case there are six points in Q (a *hexad*) that will complete an octastigm. For example: $5 + 10 = 2 + 4 + 8 + 18 + 6 + 12$ (as is easily deduced by adding the two appropriate columns of our 24×12 matrix). We get another 66 by adding the whole of Q to each of these expressions. For example, adding Q to the example we get $5 + 10 = 1 + 16 + 9 + 13 + 3 + \infty$. The permutation γ interchanges Q and N and gives another 132 octastigms, this time with two in Q and six in N.

Similarly, there are $\binom{12}{4} = 495$ ways of choosing four points in N (a *tetrad*). In each case there are eight points in Q that complete a dodecastigm. For example $5 + 10 + 20 + 17 = 1 + 4 + 16 + 9 + 18 + 3 + 6 + 12$. Adding all the points of Q to this converts it to an octastigm $5 + 10 + 20 + 17 = 2 + 8 + 13 + 0$. We have found

$$132 + 132 + 495 = 759 \text{ octastigms.}$$

How many dodecastigms? We already saw there are 495 consisting of a tetrad in N and an octad in Q. The permutation γ will give another 495 dodecastigms, with a tetrad in Q and an octad in N. There are $\binom{12}{6} = 924$ hexads in N. As we have seen, 132 of them are associated with pairs of points in Q to produce an octastigm. The other $924 - 132 = 792$ are associated with a hexad in Q to produce a dodecastigm. For example $5 + 10 + 20 + 17 + 11 + 22 = 1 + 8 + 16 + 13 + 3 + 6$. Adding Q will give another 792. Finally, remember that N and Q are themselves dodecastigms. There are altogether

$$495 + 495 + 792 + 792 + 1 + 1 = 2576 \text{ dodecastigms.}$$

No two octastigms can have five points in common; in that case adding all 16 co-ordinate columns (modulo 2) would lead to a *hexastigm*, which cannot be, because six points in a 4-space would need two, four or six of the points to be in N and, as we have seen, all these cases lead to an octastigm or a dodecastigm. Therefore, any set of five points in an octastigm determines the octastigm uniquely:

> *The* 24 *points in* PG(11, 2) *form a configuration of* 759 *octastigms; every set of five of the points belong to exactly one octastigm. The structure is a Steiner system* S(5, 8, 24), *whose symmetry group* M_{24} *is represented as a group of collineations.*

7.21 Octastigms and Dodecastigms

The special sets of eight things in S(5, 8, 24) are usually called *octads*. I have called them octastigms to emphasize their geometrical interpretation as eight gen-

Fig. 7.8 Seven tetradic points

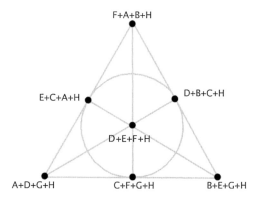

eral points in a 6-space. I have called the dodecads *dodecastigms* for the same reason. In this final section let's follow the path we took when looking at Richmond's hexastigm.

The coordinate system can be chosen so that the homogeneous coordinates of eight general points satisfy

$$A + B + C + D + E + F + G + H = 0.$$

There are $\binom{8}{2} = 28$ *duadic points* such as $A+B$ (analogous to the 15 Cremona points of a hexastigm). There are $\binom{8}{4}/2 = 35$ *tetradic points* $A + B + C + D$ etc. In the context of the previous section where the arithmetic is modulo 2, these $28 + 35 = 63$ associated points *all lie in a 5-space* because all their coordinates are linear combinations of seven, such as $A + H, B + H, C + H, \ldots, G + H$, and these seven sum to zero (modulo 2) (thus they form a 'heptastigm' in PG(5, 2)). Indeed, the 63 associated points are *all* the points of the PG(5, 2) in which they lie, because the number of points in PG(5, 2) is $1 + 2 + 2^2 + 2^3 + 2^4 + 2^5 = 2^6 - 1 = 63$.

The tetradic points associated with an octastigm in a projective space over the field F_2 lie in sevens on 30 planes (a configuration $(35_6 30_7)$ of points and planes). For example, the seven points

$$A + D + G + H = B + C + E + F$$
$$B + E + G + H = C + A + F + D$$
$$C + F + G + H = A + B + D + E$$
$$D + B + C + H = A + E + F + G$$
$$E + C + A + H = B + F + D + G$$
$$F + A + B + H = C + D + E + G$$
$$D + E + F + H = A + B + C + G$$

are all linear combinations of $A + D + G + H$, $B + E + G + H$ and $F + A + B + H$. Figure 7.8 shows the (7_3) of this PG(2, 2). A set of three disjoint octads in S(5, 8, 24)

7.21 Octastigms and Dodecastigms

is a *trio* of octads. The configuration of the 24 points in PG(11, 2) can be partitioned as a trio of octastigms in 3795 ways. Three octastigms forming a trio are given by the rows of

$$\begin{array}{cccccccc} 0 & 8 & 20 & 14 & 15 & 3 & 18 & \infty \\ 4 & 13 & 7 & 11 & 10 & 16 & 2 & 17 \\ 9 & 1 & 12 & 22 & 6 & 5 & 21 & 19 \end{array}$$

Labeling the *columns* of the array **0 1 2 3 4 5 6 ∞**, it can be shown that the actions of the α, β and γ of LF$_7$ permute the 24 points in a way that corresponds to M$_{24}$ collineations in PG(11, 2)! Each octastigm of the trio remains, of course, invariant. So we have LF$_7$ exhibited as a subgroup of M$_{24}$.

A dodecastigm is 12 general points in a 10-space. The reference system can be chosen so that their homogeneous coordinates satisfy

$$A + B + C + D + E + F + G + H + I + J + K + L = 0.$$

Associated with a dodecastigm are $\binom{12}{2} = 66$ duadic points, $\binom{12}{4} = 495$ *tetradic points* and $\binom{12}{6}/2 = 462$ *hexadic points*. In a projective space over F$_2$ these associated points all lie in a 9-space and are all the points of the PG(9, 2) in which they lie: $66 + 495 + 462 = 1023 = 2^{10} - 1$.

Consider again the generic pair of complementary dodecastigms Q and N of the configuration in PG(11, 2). The 66 duadic points of N coincide with 66 of the hexadic points of Q (and vice versa, by applying the permutation γ to the 24 points). For example, by consulting the columns of the 24 × 12 matrix on page 167 we see that **5 + 10 = 2 + 4 + 8 + 18 + 6 + 12 = 1 + 16 + 9 + 13 + 3 + ∞**. The remaining $462 - 66 = 396$ hexadic points of N coincide with hexadic points of Q. For example **5 + 10 + 20 + 17 + 11 + 22 = 1 + 8 + 16 = 13 + 3 + 6**. The tetradic points of N coincide with those of Q, for example **5 + 10 + 20 + 17 = 2 + 8 + 13 + ∞**. The associated points of two complementary dodecastigms of the 24 points in PG(11, 2) coincide!

Because all complementary pairs of dodecastigms in the configuration of 24 points in PG(11, 2) are equivalent under the M$_{24}$ collineations, we see that:

Any two complementary dodecastigms in the PG(11, 2) share the same 1023 associated points, which are all the points of a PG(9, 2).

Appendix
Notes and References

Preface

Harold Scott MacDonald Coxeter (1907–2003): English geometer who spent most of his life in Canada. An excellent biography is Siobhan Roberts' *King of Infinite Space: Donald Coxeter, the Man Who Saved Geometry*, Walker & Co. New York 2006.

Foundations: the Synthetic Approach

2. John Playfair (1748–1819): Scottish mathematician and scientist. Professor of Natural Philosophy at the University of Edinburgh.
   János Bolyai (1802–1860) and Nikolai Lobachevsky (1792–1856) discovered non-Euclidean geometry independently and almost simultaneously. Bolyai's work was published in 1832 as an appendix in a mathematical textbook written by his father. The English translation of Lobachevsky's book is *Geometrical Researches on the Theory of Parallels*, Open Court Publishing Co., Illinois 1914.
3. Donato Bramante (1444–1514): Italian architect. Bramante's most astonishing work is St. Peter's Basilica in Rome. He was also a painter, expert in the construction of correct perspective images, but left no written record of his methods. Leon Battista Alberti (1404–1472) was an architect, poet, linguist, philosopher—and mathematician. His *Della Pittura* (1435) contains the earliest systematic exposition of the methods of perspective drawing. English translation by JR Spencer: *On Painting*, Yale University Press 1956.
4. Albrecht Dürer (1471–1528): German painter and printmaker. 'A Man Drawing a Lute' is a woodcut published in the 1525 edition of Dürer's *Underweysung der Messung*.
5. Girard Desargues (1591–1661) was an engineer, architect, musician and mathematician. His *Brouillon project d'une atteinte aux evenements des rencontres du cone avec un plan* (1639) marks the beginning of projective methods in geometry. Its importance was appreciated at the time by eminent mathematicians such as Pascal and Leibniz, but thereafter was largely forgotten till the 19th century. Desargues' works were collected and published as *L'oeuvre mathématique de Desargues*, Paris, 1951 (René Taton, ed.).

11. Pappus of Alexandria (4th century): one of the last of the great Greek mathematicians. He wrote eight books, most of which have survived.
18. Some important configurations and their symmetries are lucidly discussed in Coxeter, H S M, Self-dual configurations and regular graphs, *Amer. Math. Soc. Bulletin*, **56**, 413 (1950). It was reprinted in Coxeter (1968).
18. Gino Fano (1871–1952): Italian mathematician.
26. René Descartes (1596–1650): major French philosopher and mathematician.

The Analytic Approach

38. Julius Plücker (1801–1868): German mathematician and physicist who made important contributions to analytic geometry, and early investigations of cathode rays that eventually led to the discovery of the electron.
39. Hermann Grassmann (1809–1877): German mathematician and linguist. His mathematical work was far ahead of its time and its fundamental importance was not much appreciated during his lifetime.
    Grassmann's methods were originally published in *Die lineare Ausdehnungslehre: ein neuer Zweig der Mathematik*, Wiegand, Leipzig (1844). It is discussed briefly in Coolidge 1940. English translation: Kannenberg, L, *A New Branch of Mathematics*, Open Court, Chicago 1995.

Linear Figures

58. Herbert William Richmond (1868–1948): English geometer.
    Richmond H W, On the figure of six points in a space of four dimensions, *Quart. J. Math.* **31**, 125 (1900)
    Julius Plücker (1801–1868): German geometer who introduced the idea of using six homogeneous coordinates to investigate the properties of linear figures in 3-space. Jacob Steiner (1796–1863): German geometer. A cubic surface investigated by him carries his name (see Fig. 6.1).
    Luigi Cremona (1830–1903): Italian geometer who contributed much to our understanding of algebraic curves and surfaces.
60. James Joseph Sylvester (1814–1897): major English mathematician. Founder of the American Journal of Mathematics.
65. Ludwig Schläfli (1814–1895): Swiss mathematician. One of the first geometers to explore the geometry of more than three dimensions.
68. Beniamino Segre (1903–1977): Italian geometer who made important contributions in analytic geometry and combinatorics.
    Coxeter H S M, The polytope 2_{21}, whose twenty-seven vertices correspond to the lines on the general cubic surface, *Amer. J. Math.*, **62**, 561 (1946).
72. Henry Frederick Baker (1866–1956): English mathematician. Coxeter was one of his students.
    Baker's configuration: Coxeter has employed the duads and synthemes in quite different contexts, showing how they are related to a remarkable configurations in the finite geometries PG(3, 3) and PG(5, 3): Coxeter H S M, The chords of the non-ruled quadric in PG(3, 3). *Canad. J. Math.* **10**, 484, 1958; Twelve

points in PG(5, 3) with 95040 self-transformations, *Proc. Roy. Soc. A*, **247**, 287 (1958); reprinted in Coxeter 1968.

Quadratic Figures

79. Blaise Pascal (1623–1662): French mathematician and religious philosopher. He was sixteen years old when he wrote *Essai pour les coniques* (1639), a significant work on projective geometry containing the *hexagramum mysticum* theorem.
95. Charles Julien Brianchon (1783–1864): French mathematician and chemist.
95. The 60 Pascal lines: the discovery of all these unexpected incidences is due to many eminent 19th century mathematicians—Arthur Cayley (1821–1895), George Salmon (1819–1904) and Penyngton Kirkman (1806–1905) in Britain, Julius Plücker (1801–1868) and Jacob Steiner (1796–1863) in Germany, Luigi Cremona (1830–1903) and Giuseppe Veronese (1854–1917) in Italy...
107. Felix Klein (1849–1925) proposed that all the different kinds of geometry could be classified according to the groups of transformations that preserve their structure, and emphasized the fundamental importance in geometry of the group-theoretical concept of *symmetry*. These ideas were presented in Erlangen in 1879 under the title *Vergleichende Betrachtungen über neuere geometrische Forschungen*—which became known as 'Klein's Erlangen program'. Thus a projective space is characterized by its collineation group and the various metric geometries (including the newly-discovered 'non-Euclidean' geometries) correspond to the subgroups of collineations that leave their absolute quadric invariant.
108. William Kingdon Clifford (1845–1879): English mathematician and philosopher, now best known for 'Clifford algebras'—a generalization of Hamilton's quaternions.

Cubic Figures

119. Carl Gustav Jacobi (1804–1851): German mathematician who made very many fundamental contributions.
129. The various degenerate cases of the configuration of the 27 lines lead to a classification scheme for cubic surfaces: Schläfli, L, On the distribution of surfaces of the third order into species, in reference to the presence or absence of singularities and the reality of their lines, *Phil. Trans. Roy. Soc.* **63**, 193 (1863).
131. Rudolph Clebsch (1833–1972): German mathematician who made important contributions in algebraic geometry.
    In the 19th and early 20th centuries, models demonstrating intricate geometrical concepts were sculpted in wire, wood or plaster and displayed in mathematics departments of universities. Fischer (1986) is a fine collection of photographs of these models. Many of these displays were subsequently sadly neglected—fashions in mathematics, as elsewhere, come and go. It is to be hoped that the recent advent of 'stereolithography' will lead to a revival of this delightful aspect of geometrical studies.

Quartic Figures

133. Basset (1910) is a classic on the subject of singularities of surfaces. A much more recent book on this topic is Lane, E P, *Projective Differential Geometry of Curves and Surfaces*, Porter Press 2007. Jessop (1916) deals specifically with quartic surfaces.
138. Ernst Eduard Kummer (1810–1893): German mathematician who made major contributions to number theory and hypergeometric functions.

Finite Geometries

148. Évariste Galois (1811–1832): French mathematician. He developed the foundations of the modern approach to group theory (he is the initiator of the use of the word *groupe* in this context), made significant contributions to number theory and originated the idea of finite fields—now known as Galois fields. He got himself killed in a duel before he was twenty-one.
149. Ferdinand Georg Frobenius (1849–1917): German mathematician known for important work in group theory, number theory and differential equations.
156. John Horton Conway (born 1937): English mathematician.
    Conway, J H, Three lectures on the exceptional groups, in *Finite Simple Groups* (Powell & Higman, eds.), Academic Press 1971; reprinted in Conway, J H and Sloane, N J A, *Sphere Packings, Lattices and Groups*, Springer 1988.
163. Émile Léonard Mathieu (1835–1890): French mathematician.
    The five Mathieu groups M_{11}, M_{12}, M_{22}, M_{23}, and M_{24} were the first to be discovered of the 'sporadic' finite simple groups.
164. Coxeter H S M, A symmetrical arrangement of eleven hemi-icosahedra, *Annals of Discrete Mathematics* **20**, 103 (1984).
165. Coxeter H S M, Twelve points on PG(5, 3) with 95040 self-transformations, *Proc. Roy. Soc. Lond.* A **247**, 279 (1958); Reprinted in Coxeter 1968.
168. Marcel Golay (1902–1989): Swiss mathematician and information theorist.
    The authoritative source for coding theory and finite groups is Conway & Sloane 1988.
    The configuration of 24 points in PG(11, 2) (essentially the Golay code interpreted geometrically) was discovered and investigated by John Arthur Todd (1908–1994): A note on the linear fractional groups, *J. Lond. Math. Soc.* **7**, 195 (1932); On representations of the Mathieu groups as collineation groups, *J. Lond. Math. Soc.* **34**, 406 (1959); On representations of the Mathieu group M_{24} as a collineation group, *Ann. di Math. Pure ed Appl.* **71**, 199 (1966).

A few other papers related to finite geometry that may be of interest:

Conway, J H, The miracle octad generator, *Proc. Summer School, University College Galway* (M P J Curran, ed.) Academic Press 1977.

Curtis, R T, A new combinatorial approach to M_{24}, *Mat. Proc. Cambridge Phil. Soc.* **79**, 25 (1976)

Golay, M, Notes on digital coding, *Proc. IRE*, **37**, 657 (1949).

Lord, E A, Geometry of the Mathieu groups and Golay codes, *Proc. Indian Acad. Sci. (Math. Sci.)* **98**, 153 (1988)

Mason, D R, On the construction of the Steiner system S(5, 8, 24), *J. Algebra* **47**, 77 (1977)

Paige, L J, A note on the Mathieu groups, *Can. J. Math.* **9**, 15 (1956)

Bibliography

In addition to the works mentioned in the Appendix, here are some books that you may find useful for supplementary reading if you want to delve deeper. This is a highly personal selection—simply a few books I found particularly stimulating while learning about this fascinating subject. They are historic treasures; some of them have been recently reprinted and a few of them can be downloaded free from the internet.

Baker, H F, *Principles of Geometry* (6 Vols.), Cambridge University Press, Cambridge, 1940.
Baker, H F, *A Locus with 25920 Linear Self-Transformations*, Cambridge University Press, Cambridge, 1946.
Basset, A B, *A Treatise on the Geometry of Surfaces*, Cambridge University Press, Cambridge, 1910.
Bennett, M K, *Affine and Projective Geometry*, Wiley, New York, 1995.
Coxeter, H S M, *The Real Projective Plane* (2nd edn.), Cambridge University Press, Cambridge, 1955.
Coxeter, H S M, *Introduction to Geometry*, Wiley, New York, 1961a.
Coxeter, H S M, *Non-Euclidean Geometry* (4th edn.), University of Toronto Press, Toronto, 1961b.
Coxeter, H S M, *Twelve Geometric Essays*, Southern Illinois University Press, Illinois, 1968.
Cremona, L, *Elements of Projective Geometry*, Oxford University Press, Oxford, 1885 (tr. C Leudesdorf).
Conway, J H & Sloane, N J A, *Sphere Packings, Lattices and Groups*, Springer, Berlin, 1988.
Coolidge, J L, *A History of Geometrical Methods*, Oxford University Press, Oxford, 1940. Dover reprint 1963.
Coolidge, J L, *A History of the Conic Sections and Quadric Surfaces*, Oxford University Press, Oxford, 1945.
Fischer, G, *Mathematische Modelle: Mathematical Models*, Vieweg, Braunschweig/Wiesbaden, 1986.

Henderson, A, *The Twenty-Seven Lines upon the Cubic Surface*, Cambridge University Press, Cambridge, 1911.

Hilbert, D, *The Foundations of Geometry*, Open Court Publishing Co., Illinois, 1950 (English translation of Hilbert's doctoral dissertation *Grundlagen der Geometrie* 1899).

Hilbert, D & Cohn-Vossen, S, *Geometry and the Imagination*, Chelsea Publishing Company, New York, 1952 (originally *Anschaulische Geometrie*, Springer 1932).

Hirschfeld, J W P, *Projective Geometries over Finite Fields*, Oxford University Press, Oxford, 1979.

Hudson, J W H T, *Kummer's Quartic Surface*, Cambridge University Press, Cambridge, 1905.

Jessop, C M, *Quartic Surfaces with Singular Points*, Cambridge University Press, Cambridge, 1916.

Klein, F, *Elementary Mathematics from an Advanced Standpoint*, Geometry, Macmillan, Vol 1, (originally *Elemantarmathematik vom Höherem Standpunkt aus; Band 1: Geometrie*, Leipzig 1908).

Pedoe, D, *An Introduction to Projective Geometry*, Pergamon, Elmsford, 1963.

Salmon, G, *A Treatise on the Analytic Geometry of Three Dimensions* (4th edn.), Hodge, Figgis & Co, Dublin (1882).

Segre, B, *The Non-Singular Cubic Surfaces*, Oxford University Press, Oxford, 1942.

Semple, J G & Kneebone, G T, *Algebraic Projective Geometry* (2nd edn.), Oxford University Press, Oxford (1952).

Todd, J A, *Projective and Analytic Geometry*, Pitman, London, 1947.

Veblen, O & Young, J M, *Projective Geometry* (2 Vols.), Ginn & Co., Needham Heights, 1918

Index

Symbols
16 point theorem, 106

A
Absolute conic, 89
Absolute quadric, 109
Affine classification of conics, 81
Affine classification of quadrics, 103
Affine cubics, 120
Affine geometry, 10, 36, 47, 120
Alberti, 3, 173
Algebraic geometry, 133
Algebraic surface, 125
Algebraic variety, 133
Alternating symbol, 39
Architects, 1
Associated desmic systems, 138
Associated lines, 65
Associated trihedron pair, 66
Associativity, 16, 24
Asymptotic cone, 104
Axioms, 2, 10, 24

B
Baker, 72, 174
Binary Golay code, 168
Binode, 133
Bolyai, 2, 173
Bramante, 3, 173
Brianchon, 95, 175
Burkhardt's primal, 76

C
Canonical form for a collineation, 32
Canonical form for a conic, 83
Canonical form for a desmic system, 69
Canonical form for a plane cubic, 117
Canonical form for a quadric, 99
Carnevale, 3
Cartesian coordinates, 26
Cayley, 175
Cayley lines, 98
Circle, 1, 49, 87, 92
Civa, 48
Clebsch, 175
Clebsch surface, 131, 132
Clifford, 175
Clifford parallels, 108
Collineation, 31, 38, 47, 71, 110, 140, 146, 160
Commutativity, 15, 16
Complete quadrangle, 18
Complete quadrilateral, 7, 18, 51, 69, 87, 146
Complex line, 48, 103, 120
Configurations, 17
Conic at infinity, 89
Conic envelope, 82
Conic sections, 81
Conics, 79
Conway, 156, 176
Coxeter, 163, 173, 174
Coxeter's configuration, 163
Cremona, 174, 175
Cremona point, 61, 62
Cremona points, 58, 60
Cross-ratio, 45, 47
Cubic curve, 115, 120
Cubic surface, 125
Cusp, 115, 133
Cuspidal edge, 123

D
Degenerate conic, 80, 116
Degenerate cubic curve, 116, 125, 131
Degenerate quartic curve, 109, 124

Degenerate quartic surface, 136
Desargues, 5, 173
Desargues' configuration, 7, 8, 18, 22, 55, 89, 98
Desargues' theorem, 5, 11, 13, 34, 56
Desargues' theorem extensions, 63, 64
Descartes, 26, 174
Desmic surface, 137
Desmic system, 69
Desmic tetrahedra, 70
Determinant, 29, 80, 93, 117, 125, 128, 147
Determinantal canonical form, 128
Diagonal point, 8, 51, 54
Diagonal triangle, 8, 20, 69, 86
Dimensionality, 25, 30, 36, 145
Direct isometry, 111
Discriminant, 44, 80
Dodecastigm, 168
Double point, 116, 131
Double-six, 65, 73, 131
Duadic points, 170
Duads, 7, 18, 55–57, 60, 62, 158
Dual coordinates, 39, 40
Duality, 5, 6, 19, 25
Dürer, 4, 173

E

Eckhardt points, 131
Eigenvalue, 32
Eigenvector, 32
Ellipse, 81
Ellipsoid, 103
Elliptic geometry, 91, 107
Elliptic hyperboloid, 99
Elliptic paraboloid, 103
Equianharmonic points, 50, 72
Erlangen program, 175
Euclid, 1
Euclidean geometry, 2, 26, 91, 107
Exceptional groups, 156
Exceptional points, 129

F

Fano, 174
Fano's configuration, 18, 146, 161
Fermat's 'little' theorem, 155
Field, 26, 148
Finite geometries, 145
Fixed line, 32
Fixed point, 32
Frobenius, 176
Frobenius automorphism, 149, 155

G

Gallucci's theorem, 106
Galois, 176
Galois fields, 148
Generators, 99, 162
Golay, 176
Golay codes, 168
Golden rectangle, 158
Grassmann, 174
Grassmann coordinates, 39

H

Hamiltonian circuit, 20
Hamming code, 168
Harmonic conjugate, 8, 45, 51, 53, 59, 69, 90
Harmonic construction, 9
Harmonic points, 51, 55, 58, 59
Harmonic range, 8, 51
Hemi-icosahedron, 163
Hermitian conjugate, 112
Hermitian matrix, 49
Hessian, 119, 134
Hexacode, 168
Hexad, 163
Hexadic points, 171
Hexagrammum mysticum, 92
Hexastigm, 58, 61–63, 74, 165, 166, 169, 170
Homogeneous coordinates, 27
Homographic parameter, 43, 53, 84, 122
Homographic parameter on a conic, 84
Homographic parameter on a twisted cubic, 122
Homography, 44, 49, 111, 160
Hyperbola, 81, 120, 138, 144
Hyperbolic geometry, 91
Hyperbolic paraboloid, 104
Hyperplane, 30, 37, 38, 47, 53, 74, 128, 132, 154

I

Icosahedron, 53, 158, 159, 163
Infinity, 10, 11, 14, 15, 29, 36, 47, 71, 81, 88, 89, 91, 94, 103, 108, 112, 120, 147, 162
Inflection, 115, 119, 150
Inflectional tangent, 119
Invariant line, 32
Involution, 46
Involutory hexad, 46
Isotropic lines, 89, 91, 92, 107

J

Jacobi, 175
Jacobi configuration, 120, 147, 150, 151

Index

K
Kirkman, 175
Kirkman points, 98
Klein, 175
Kummer, 176
Kummer's surface, 138

L
Left parallel, 109, 110
Left translation, 111
Levi graph, 19–24, 61, 98
Line at infinity, 10, 11, 29, 36, 47, 48, 81, 88, 91, 94, 120, 125, 147
Linear asymptotes, 120
Linear fractional group, 146, 155, 156
Lobachevsky, 91

M
Mathieu, 176
Mathieu group, 163, 166–168
Menelaus, 48
Metric planes, 89
Metric spaces, 107
Mutually inscribed pentagons, 23
Mutually polar triangles, 88
Möbius configuration, 161

N
N-dimensional axioms, 24
Net of quadrics, 124
Node, 73, 77, 115, 116, 118, 127, 133, 135, 138, 140, 141
Non-Desarguesian, 6
Non-Euclidean, 2, 91

O
Octad, 163
Octastigm, 169
Opposite isometry, 110
Orthogonal axes, 26
Orthogonal group, 110
Orthogonality, 26, 90
Osculating plane, 123, 124

P
Pappus, 174
Pappus configuration, 20
Pappus's theorem, 12, 33
Parabola, 81
Parabolic asymptote, 120
Parallel, 2, 4, 10, 11, 16, 26, 36, 47, 88–91, 94, 108, 142, 147
Parametric form for a conic, 83
Parametric form for a plane cubic curve, 118
Parametric form for a twisted cubic, 121
Pascal, 175
Pascal lines, 62, 95
Pascal's theorem, 92, 100
Pascal's triangle, 54, 151
Pencil of cubics, 119
Pencil of planes, 122
Pentahedron, 128, 135, 136
Permutations, 18, 40, 50, 56, 60, 62, 63, 72, 76, 89, 95, 131, 140, 156, 160, 164, 169
Perspective drawing, 3
Perspective triangles, 5
Perspectivity, 5, 15, 84, 94, 150
Plane at infinity, 11, 36, 47, 71, 103, 104, 107, 108, 138, 162
Plane cubic curve, 115
Plücker, 174, 175
Plücker coordinates, 38
Plücker lines, 58, 62, 97
Polar conic, 115, 116, 119, 120
Polar line, 85, 115
Polar plane, 102, 127
Polar quadric, 127, 134, 135
Polarity with respect to a conic, 85
Polarity with respect to a cubic curve, 115
Polarity with respect to a quadric, 102
Polytope, 68
Postulates, 1
Projective equivalence, 28, 29, 31, 36, 71, 81, 84, 121, 137, 146
Projective line, 43, 48, 49
Projective transformation, 31, 43, 46, 76, 146
Pseudo-Euclidean geometry, 92
Pseudo-orthogonal group, 110

Q
Quadrangle, 18, 20, 46, 50, 51, 69, 86, 87, 146
Quadratic, 48, 79, 115, 123, 148
Quadratic form, 49
Quadratic residues, 162, 164, 165, 167, 168
Quadric, 99
Quadric at infinity, 107
Quadric cone, 100, 125, 127, 133
Quadrilateral, 7, 18, 51, 69, 87, 146

R
Reference simplex, 59, 74, 154
Reference tetrahedron, 30, 31, 71, 124
Reference triangle, 28, 32, 34, 45–47, 52, 82, 83, 88, 89, 92, 94, 117
Regular configuration, 19
Richmond, 174
Richmond's hexastigm, 58
Riemann, 91

Right parallel, 109, 110
Right translation, 111

S
Salmon, 175
Salmon points, 98
Schläfli, 174
Schläfli's double-six, 65
Segre, 68, 174
Self-duality, 19, 20, 22, 53, 61
Self-inscribed decagon, 23
Self-inscribed heptagon, 20, 161
Self-inscribed nonagon, 20
Self-polar triangle, 86
Simplex, 53, 73, 151
Singularity, 123, 133
Skew lines, 68, 104
Skewsymmetry, 39
Special linear group, 31, 155
Special unitary group, 112
Steiner, 174, 175
Steiner planes, 58
Steiner points, 96
Steiner system, 163
Steiner's surface, 134
Summation convention, 37, 115, 125
Sylvester, 174
Sylvester canonical form, 128, 136
Sylvester lines, 60, 72
Sylvester's duads and synthemes, 60, 158

Symmetric group, 62
Symmetric matrix, 49, 80, 82, 99
Symmetry group, 20, 56, 60, 72, 77, 140, 141, 155, 161, 163, 164, 166, 167, 169
Synthemes, 60, 158

T
Tangent, 81, 89, 102, 115, 116, 120, 123
Tangent cone, 127, 133, 134
Tangent plane, 101, 107, 123
Translation, 111
Transversal, 104
Trihedron, 56, 66, 129
Triple point, 116
Tritangent plane, 66, 125
Tutte-Coxeter graph, 61
Twenty-seven lines, 64, 129

U
Unitary matrix, 112

V
Vanishing point, 4
Vector space, 36
Veronese, 175
Virtual conic, 83, 103
Virtual quadric, 99

W
Whitney's umbrella, 134